Foreword

KV-638-413

Description of work

The Research Project, RP523, which led to this Report, was undertaken in two parts, which have been brought together in an integrated Report.

Part 1 reviewed the legislative and other environmental requirements which have driven the development of 'compliant' coatings for construction steelwork forward rapidly in the last five years. It also reviewed the generic options available currently to meet these requirements. In particular, it reviewed the *Environmental Protection Act 1990* (EPA) and associated guidance. This was achieved by a literature review and by consultation with environmental specialists, design professionals, manufacturers of high-performance coatings, steelwork fabricators, painters and end-users generally. A very wide range of viewpoints was sought and has been represented accordingly.

Part 2 developed design guidance on the use of 'compliant' coatings for the corrosion protection of construction steelwork for designers and specifiers. It also surveyed good practice.

The work was carried out by joint contractors: Arup Research and Development and the Paint Research Association (PRA).

The project was developed and managed by Mrs Ann Alderson, Research Manager at CIRIA, with advice and guidance from a Project Steering Group whose support and valuable contributions are gratefully acknowledged. The Steering Group comprised:

Chairman

Dr Chris Sketchley	Scott Wilson Kirkpatrick & Co Ltd
Mr Peter Allen	British Constructional Steelwork Association
Mr John Caves	London Underground Ltd
Eur Ing Simon Clarke	Sandberg Consulting Engineers
Mr Lou Fear/Dr Jim Breakell	Mott MacDonald Special Services
Mr Chris Hitchen	Defence Research Agency, Structural Materials Centre
Mr Roger Hudson	British Steel plc
Mr Awtar Jandu	Highways Agency
Mr Douglas Lewis	BCF High Performance Coatings Group
Dr Peter Morris	W & J Leigh & Co
Mr Colin Thompson	Northumbrian Water *representing the Water Utilities*
Mr Roy Thurgood	Construction Innovation and Research Management Division, Department of the Environment
Mr Iain Wesley	W S Atkins Structural Engineering
Mr Hugh Williams	British Coatings Federation Ltd

CIRIA is also grateful to Mr Simon Smith, Air and Environmental Quality Division of the Department of the Environment, for his input on the proposed revisions to current legislation.

The project was funded by the Department of the Environment, Construction Directorate, and CIRIA's Core Programme.

Report 174

UNIVERSITY OF STRATHCLYDE
30125 00555907 4

1997

New paint systems for the protection of construction steelwork

Graham Gedge
Dr Nigel Whitehouse

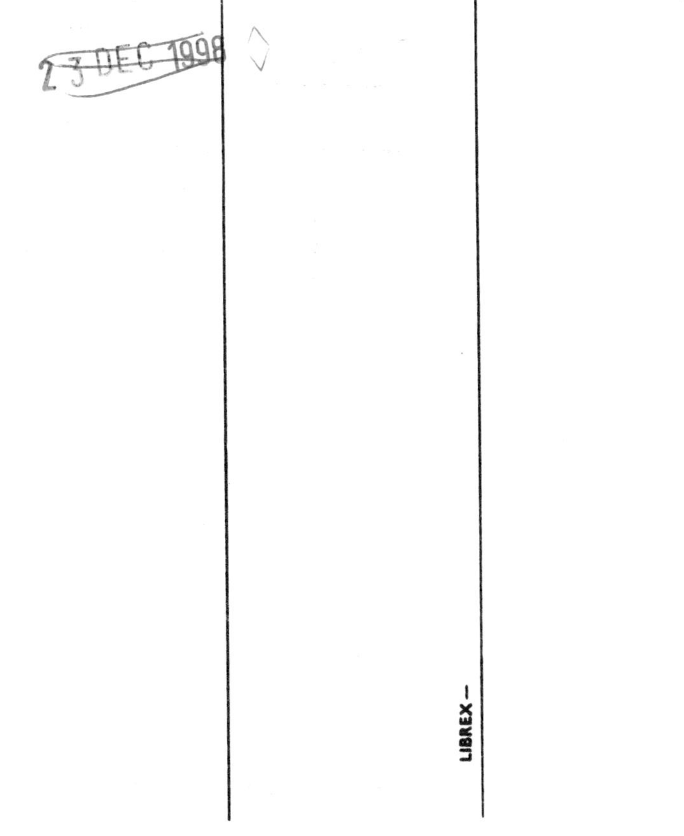

CONSTRUCTION INDUSTRY F
6 Storey's Gate, Westminster, L
E-mail switchboard @ ciria.org.uk
Tel 0171-222 8891 Fax 0171-222 1708

Summary

This report gives guidance on the selection, application and specification of coatings for use in the general steel fabrication industry. It considers the impact of *The Environmental Protection Act (1990)* on the coatings industry and offers guidance on the use and specification of coating materials that comply with the requirements of the Act.

The topics covered include a review of the relevant Environmental, Health and Safety legislation, the range of available materials, the relevance and use of these in general steel construction, the advantages and disadvantages of the new materials, including cost information. The report also includes standard material specifications for a range of environments commonly encountered in general construction. These materials specifications can be incorporated in existing specifications, notably the SCI/BCSA *National Steelwork Specification for Building Construction (NSSS)*.

Graham Gedge, Dr Nigel Whitehouse
New paint systems for the protection of construction steelwork
Construction Industry Research and Information Association
CIRIA Report 174, 1997

© CIRIA 1997

ISBN: 0 86017 472 7

ISSN: 0305 408 X

Published by CIRIA, 6 Storey's Gate, Westminster, London SW1P 3AU. All rights reserved. With the exception of the CIRIA Specifications E1-E2 and I1-I5 no part of this publication may be reproduced or transmitted in any form or by any means, including photocopying and recording, without the written permission of the copyright holder, application for which should be addressed to the publisher. Such written permission must also be obtained before any part of this publication is stored in a retrieval system of any nature.

Keywords

Environmental, specification, volatile organic compounds (VOCs) Coating, compliant, paint, corrosion protection, steelwork, paint systems

Reader interest	Classification	
Designers and Specifiers of general construction steelwork.	AVAILABILITY	Unrestricted
	CONTENT	Review of current legislation; guidance on specification
	STATUS	Committee guided
	USER	Designers and specifiers of general construction steelwork. Steel producers, fabricators, applicators and coating manufacturers.

D
624.1821
GED

Contents

List of Figures

List of Tables

Glossary

SELECTED GLOSSARY OF PAINT AND RELATED TERMS

NOTE:	A comprehensive Glossary of paint and related terms is given in *BS2015*: 1992 and *BS EN 971-1*: 1996.
ALKYD RESIN	A synthetic resin made by condensation between a polyhydric alcohol such as glycerol, and a polybasic acid such as phthalic acid (normally in the form of the anhydride).
ANTI-CORROSIVE PAINT	A coating material used to retard the corrosion of metals and, more particularly, specially formulated to retard the rusting of iron and steel.
BARRIER COAT	A coating material used to isolate a coating system from the substrate to which it is applied in order to prevent chemical or physical interaction.
BINDER	The non-volatile part of the medium which forms the film.
BLAST PRIMER/ PRE-FABRICATION PRIMER	A coating material that is applied to a ferrous substrate directly after blast cleaning.
COATING MATERIAL	A product, in liquid or in paste or powder form, that, when applied to a substrate, forms a film possessing protective, decorative and/or other specific properties.
COATING SYSTEM	The sum total of the coats of coating materials which are to be applied or which have been applied to a substrate.
CROSS-LINKING AGENT	A compound that will react chemically with a polymeric material, giving rise to a three-dimensional network which is usually insoluble in common solvents.
CURING	The process of condensation or polymerization of a material by heat or chemical means resulting in the full development of the desired properties.
CURING AGENT	An additive that promotes the chemical curing of a film.
DRY TO HANDLE	The state of drying when a coat can be handled without damage.
EPOXY RESIN	A synthetic resin containing epoxide, in which the final polymer is formed as a result of a reaction taking place substantially in the epoxide groups.

ETCH PRIMER/WASH/ PRE-TREATMENT PRIMER/SELF-ETCH PRIMER	A coating material often supplied as two separate components that are mixed immediately prior to application and have a limited pot-life. The mixed coating material contains balanced proportions of a chromate-based inhibitive pigment, phosphoric acid and a synthetic resin binder in a mixed alcohol solvent, generally a polyvinyl butyral.
EXTENDER	A material in granular or powder form, practically insoluble in the application medium and used as a constituent of paints to modify or influence certain physical properties.
FILM	A continuous layer resulting from the application of one or more coats to a substrate.
FILM FORMATION	The process by which coating materials, when applied to a substrate, are transferred into a cohesive layer.
FINISH	The final or only coat in a coating system.
HIGH SOLIDS	A term applied to coating materials in which, by the choice of suitable ingredients, the content of volatiles present is kept to a minimum, consistent with the maintenance of satisfactory application properties.
HIGH-BUILD	The property of a coating material which permits the application of a coat of greater than normal thickness.
INHIBITIVE PIGMENT	A pigment that retards or prevents corrosion of metals by chemical and/or electrochemical means, as opposed to acting purely as a barrier.
ISOCYANATES	A class of organic compounds, embodying an isocyanate (NCO) group that reacts with polyesters and polyethers to form polyurethane resins.
LATEX	Originally a natural rubber latex; now also applied to dispersions of various synthetic resins.
MEDIUM	The sum total of the constituents of the liquid phase of a coating material.
PIGMENT	A substance, generally in the form of fine particles, which is practically insoluble in media and which is used because of its optical, protective or decorative properties. NOTE: In particular cases, for example corrosion inhibiting pigments, a certain degree of water solubility is necessary.
POLYURETHANE RESIN	A synthetic resin produced by reacting a polyhydroxyl reactant, normally of polyester or polyether structure, with a polyisocyanate.
POT LIFE	The maximum time during which a coating material supplied as separate components should be used after they have been mixed together.
PRIMER/FINISH	A single-coat application that attempts to combine the properties of both a primer and a finish.

PRIMER/FINISH	A specially formulated coating material used to prime and finish simultaneously a suitably prepared substrate.
RESIN	A synthetic or naturally high molecular weight material with film-forming properties.
SEALER	A clear or pigmented liquid applied to absorbent substrates prior to painting which when dry reduces the absorptive capacity of the substrate.
SILICONE-MODIFIED RESIN	A synthetic resin, the properties of which have been modified with silicone – usually to improve resistance to weathering or to heat.
SOLVENT	A liquid, usually volatile, that is used in the manufacture of a coating material to dissolve or disperse the constituents responsible for film formation and that evaporates during drying and therefore does not become part of the dried film.
SOLVENT-BORNE MATERIAL	A material that is dispersed or dissolved in suitable organic solvents.
SOLVENT-FREE PAINT	An organic coating material containing no volatile thinner.
STRIPE COAT	An extra cost of paint applied locally to areas where the shape and/or the plane of application may result in thin coatings, e.g. at edges. Stripe coats are usually applied first of all and are then covered by a full coat. Stripe coating ensures a double thickness of paint on vulnerable areas.
THIXOTROPIC AGENT	A chemical substance which, when added to a coating material in small quantities, introduces thioxotropy.
THIXOTROPY	The process whereby a coating material undergoes a reduction in body when disturbed mechanically and reverts slowly to the original condition on standing.
TIE COAT	A product, usually unpigmented, designed to improve intercoat adhesion by slightly softening the dry film to which it is applied and being softened in turn by the coating material subsequently applied to it.
TOUCH DRY	A state of drying when slight pressure applied by a finger does not have an imprint or reveal tackiness.
TWO-PACK PAINT	A coating material that is supplied in two parts which have to be mixed in the correct proportions before use. The mixture will then remain in a usable condition for a limited time.
VOLATILE ORGANIC COMPOUND	Fundamentally, any organic liquid and/or solid that evaporates spontaneously at the prevailing temperature and pressure of the atmosphere with which it is in contact.
WATER-BORNE PAINT	A paint in which the pigment and binder are dispersed or dissolved in a continuous phase that consists essentially of water. [Some users prefer the term 'water-based', but this Report follows current convention.]

ZINC DUST	Finely divided zinc metal used as a pigment in protective coating materials for iron and steel.
ZINC SILICATE PRIMER	An anti-corrosive coating material for iron and steel based on a silicate resin and incorporating zinc dust in a concentration sufficient to protect the substrate cathodically.
ZINC-RICH PRIMER	An anti-corrosive coating material for iron and steel, incorporating zinc dust in a concentration sufficient to give cathodic protection, whereby the dry film is electrically conductive, enabling the zinc metal to corrode preferentially to the substrate.

Acronyms

ACOP	Approved Code of Practice (UK)
APC	Air Pollution Control (UK)
AIM	Architectural and Industrial Maintenance (USA)
BATNEEC	Best Available Technique (or Technology) Not Entailing Excessive Cost
BCF	British Coatings Federation
BPM	Best Practicable Means
BSI	British Standards Institution
CAA	Clean Air Act (USA)
CHIP	Chemicals (Hazard Information and Packaging) Regulations, (UK)
CIRIA	Construction Industry Research and Information Association (UK)
CDM	Construction (Design and Management) Regulations, (UK)
CEN	European Standardisation Committee
CEPE	European Confederation of Paint, Printing Ink and Artists' Colours Manufacturers' Associations
CHIP	Chemicals (Hazard Information and Packaging) Regulations, (UK)
CONIAC	Construction Industry Advisory Committee (UK)
COPA	Control of Pollution Act (UK)
COSHH	Control of Substances Hazardous to Health (Regulations, UK)
CP	Code of Practice
CRINE	Cost Reduction In the New Era
DD	Draft for Development
DFT	Dry Film Thickness
DIS	Draft International Standard
DOE	Department of the Environment (UK)
DPC	Draft for Public Comment
EC	European Community
ECISS	European Committee for Iron and Steel Standardisation
EEC	European Economic Community
EH	Environmental Hygiene
EN	European Standard
ENV	European Pre-Standard
EPA	Environmental Protection Act (UK)
EU	European Union
FDIS	Final Draft International Standard
FIP	Federal Implementation Plan (USA)
FIPEC	French Federation of the Paints, Inks, Colours, Glues, and Adhesives Industries
GG	General Guidance Note (UK)

HA	Highways Agency (UK)
HAP	Hazardous Air Pollutant (USA)
HSWA	Health and Safety at Work Act (UK)
HMIP	Her Majesty's Inspectorate of Pollution (UK)
HMSO	Her Majesty's Stationary Office
HSC	Health and Safety Commission (UK)
HSDS	Health and Safety Data Sheet
HSE	Health and Safety Executive (UK)
IPC	Integrated Pollution Control (UK)
IPPC	Integrated Pollution Prevention and Control (EU)
ISO	International Standards Organisation
MIO	Micaceous Iron Oxide
MOD	Ministry of Defence (UK)
MSDS	Material Safety Data Sheet
NAAQS	National Ambient Air Quality Standard (USA)
NESHAP	National Emission Standard for Hazardous Air Pollutants (USA)
NPCA	National Paint and Coatings Association (USA)
NRA	National Rivers Authority
NSSS	National Structural Steelwork Specification for Building Construction
OEL	Occupational Exposure Limit
PG	Process Guidance Note, (UK)
PPE	Personal Protective Equipment
PRA	Paint Research Association (UK)
PVB	Polyvinyl Butyral
QA	Quality Assurance
RH	Relative Humidity
SEPA	Scottish Environment Protection Agency (UK)
SI	Statutory Instrument (UK)
SIP	State Implementation Plan (USA)
TA Luft	German Air Pollution Control Regulations
UK	United Kingdom
UNECE	United Nations Economic Commission for Europe
USA	United States of America
VOC	Volatile Organic Compound
WMP	Waste Management Paper (UK)
WRA	Waste Regulation Authority (UK)

1 Introduction

1.1 BACKGROUND

Traditionally, construction steelwork has been protected with high performance, solvent-borne paint systems consisting, typically, of an anti-corrosive primer, one or two intermediate coats, and an appropriate finish. Continuing concern over the uncontrolled release of volatile organic compounds (VOCs) into the atmosphere, and the adverse effect which such compounds are known to have on the environment, has led to increasingly restrictive legislative controls in most developed countries.

In the UK, the EPA has established a radical regime of pollution control extending to air, water, and land. *Process Guidance Notes* (PGs) have been issued in support of the Act for a range of manufacturing operations.

The relevant Process Guidance Note for the *Coating of Metal and Plastic* is *PG6/23(97)*. This Note is directed towards the application of coatings in enclosed areas, such as fabrication shops and painting halls. It provides two alternative approaches to the control of VOC emissions :

- The installation of abatement equipment to remove organic solvents from atmospheric emissions.
- The use of 'compliant' coatings with reduced (defined) organic solvent contents.

The paint industry, in support of the statutory control regimes which the EPA legislation has introduced, has developed a comprehensive range of 'compliant' coatings. Almost all steel fabricators and their associated painting contractors, have opted to comply with legislation by introducing these new products. Very few have chosen to install abatement equipment.

'Compliant' coatings are not covered by *BS 5493 : 1977*, the current *Code of Practice for the Protective Coating of Iron and Steel Structures against Corrosion*. 'Low VOC paint systems' will, however, be included in the examples listed in Annex A of *BS EN ISO 12944-5 (Paints and Varnishes – Corrosion Protection of Steel Structures by Protective Paint Systems – Part 5 – Protective Paint Systems)*, a new Standard (in eight parts) which is being developed currently by ISO.

In the absence of relevant standards, many designers, specifiers, and clients understandably have been reluctant to change from proven specifications. This Report seeks to provide independent, authoritative good practice guidance for designers and specifiers on the performance requirements, use, selection, and specification of compliant paint systems.

It culminates in a set of standard materials specifications (see Section 10).

This CIRIA Report relates to general (ungalvanised) steelwork fabrication for environments normally found in building and civil engineering (*e.g.* specified in accordance with NSSS). It reflects UK legislation, its application in good practice and available materials, as at March 1997. Readers/users should note that the legislation and materials are still being developed, and will change with time.

1.2 COMPLIANT COATINGS AND CURRENT TRENDS

The term compliant coating is defined in *PG6/23* (see Section 3.5.3) in terms of solvent content. In this Report, the term is used in a more general way to cover coatings that are generally regarded as 'user friendly', but does not necessarily imply the material complies with *all* legislation relevant to a given situation, only that the VOC limits given in *PG6/23* are met.

In terms of a contract specification, compliant coatings and their application do not differ significantly from the materials they replace. Compliant coatings do differ significantly, however, in important aspects relating to composition, thickness and available materials. The advantages of compliant materials to the specifier are not immediately apparent: few specifications have required that compliant coatings be used. This must change, but will only do so if the specifiers have confidence in the performance of the available materials and, to a lesser extent, that the materials do not cost significantly more, in terms of applied costs, than the materials they replace.

The introduction of the EPA could be viewed as yet another unnecessary legislative burden imposed on the industry, inevitably leading to increased costs. However, there are good reasons to believe that this pessimistic outlook is wrong. The EPA could result in considerable benefit to the industry.

In addition, it is recognised that coating manufacturers continue to develop new products, some of which might be fundamentally different from those now available. Where such products are known, their possible impact on the construction industry is discussed, although specific guidance on their use is not attempted here.

1.3 SCOPE OF THE REPORT

1.3.1 Principal scope

The Report considers compliant coating specifications for corrosion prevention of new construction steelwork in various internal and external environments normally found in buildings and structures in the UK. It does not cover the refurbishment and maintenance of existing structures. It is specifically intended for use in general steel fabrication, as opposed to more specialist applications. It is aimed at the immediate impact of the 1996 requirements of the EPA, but also takes account of the proposed 1998 requirements.

For guidance, this Report defines 'general fabrication' contracts as those that are typically controlled by the *National Structural Steelwork Specification for Building Construction (NSSS)*. Examples of such structures are: speculative office buildings, retail outlets, warehouses and similar structures.

This CIRIA Report relates to general (ungalvanised) steelwork fabrication for environments normally found in building and civil engineering (*e.g.* specified in accordance with NSSS). It reflects UK legislation, its application in good practice and available materials, as at March 1997. Readers/users should note that the legislation and materials are still being developed, and will change with time.

While a similar approach and specifications may be adopted on other structures, the guidance given will need to be augmented by consideration of the specific needs of the particular project.

The Report **does not** consider the protection of structures in special environments such as offshore or maritime applications, the petrochemical or process industries, where more severe environments are encountered. Nor does it cover structures commissioned or maintained on behalf of major utilities with their own specifications and/or performance requirements which may differ from those given in this document.

The Report **does not** deal with the application of coating systems where steel has previously been hot dip galvanised.

Specifically, the report:

- Identifies and reviews relevant environment and health and safety legislation, particularly *The Environment Protection Act (Part 1)* (EPA), *Control of Substance Hazardous to Health* (COSHH), and *The Construction (Design and Management)* (CDM) *Regulations* in respect of the use of protective coatings for construction steelwork.
- Identifies the **generic** types of coating that are currently available and appropriate to use in the construction industry.
- Offers good practice guidance on the selection and specification of compliant coatings, based on currently available raw materials.
- Includes suggested default specifications for use in a wide range of conditions that are commonly encountered in UK construction.

1.3.2 Good practice

It is assumed that the reader/specifier typically is not a coatings specialist and will rely on:

- general good practice in the use of coatings and corrosion protection systems, including appropriate British and ISO Standards
- this Report, in particular the standard material specifications it provides (see Section 10)
- the manufacturer's instructions for the product(s) actually used.

Particular emphasis is given to the importance of seeking and obtaining the manufacturer's instructions and guidance on specific products, in view of the novelty of the current and emerging compliant coating systems.

On this basis, designers/specifiers are most likely to ensure a satisfactory product and the desired performance. In addition, should a dispute arise as to the quality of work, the building owner and/or specifier is likely to have the most appropriate basis for any remedy – namely, as a consumer.

The Report reflects UK legislation, its application in good practice and available materials, as at March 1997. Readers and users should note that the legislation and the

---

This CIRIA Report relates to general (ungalvanised) steelwork fabrication for environments normally found in building and civil engineering (*e.g.* specified in accordance with NSSS). It reflects UK legislation, its application in good practice and available materials, as at March 1997. Readers/users should note that the legislation and materials are still being developed, and will change with time.

materials are still being developed, and will change with time, but an overview begins in Section 3.

'Good practice' is taken to be that established in current general guidance (*e.g.* the NSSS) and the CIRIA specifications given in Section 10. It is also reflected in the incorporated lessons from recent consultations with specifiers, fabricators and other clients and users conducted as part of the research for the Report.

1.3.3 Further information

The Report provides background information on environmental legislation in the USA and the more environmentally-aware member states of the European Union (EU). It also reviews work within standards organisations which is relevant to compliant coatings and lists other relevant technical publications in the current literature.

2 An introduction to paint technology and protective coatings

2.1 GENERAL

Construction steelwork is usually protected from corrosion by two or more coats of paint (*i.e.* a protective coating system). For new construction, the paints are applied at an appropriate stage in the fabrication process. Surface preparation and the application of a primer and intermediate coats is usually carried out in the fabrication shop or at the works of a specialist painting sub-contractor. The finishing coat is usually applied on site after the steelwork has been erected and any erection damage to the shop-applied coats has been repaired. The responsibility for coating application, therefore, rests with the fabricator. Most protective coating specifications consist of the following functional components:

- an anti-corrosive primer
- intermediate (or build) coats
- a protective/decorative finish (or topcoat).

Each element of the paint system has a particular function and the omission of any part may have a detrimental effect on protection and durability.

2.2 THE CONSTITUENTS OF PAINTS

Paint consists of a medium (the liquid part of the coating), pigments, extenders, and various additives in minor amounts. The medium comprises the film-forming resin and the solvent. Solvents may be organic liquids or water. It is convenient to divide the many different raw materials which may go into a paint into groups, according to their function in the coating:

BINDERS – The binder, or resin, is the non-volatile film-forming part of the medium. It ensures adhesion to the substrate and cohesion within the paint film. Film formation is determined by the binder type which also influences greatly the film strength and other physical and chemical properties. Typical binders are alkyds, epoxies, and polyurethanes.

SOLVENTS – Solvents reduce the viscosity of the binder, allowing the paint to be made and, later, applied. They form the volatile part of the medium and should evaporate during application and film formation. The choice of solvent influences drying time and flash point.

PIGMENTS – Pigments determine the colour and opacity of a paint. In protective coatings, however, they may introduce other properties such as corrosion protection

This CIRIA Report relates to general (ungalvanised) steelwork fabrication for environments normally found in building and civil engineering (*e.g.* specified in accordance with NSSS). It reflects UK legislation, its application in good practice and available materials, as at March 1997. Readers/users should note that the legislation and materials are still being developed, and will change with time.

and fire retardancy. Pigments also have a major effect on physical properties of paints such as film hardness and water resistance. Micaceous iron oxide (MIO) is a functional pigment often used in intermediate coats of protective paint systems. It is a lamella pigment which orientates itself in the film like slates on a roof. In so doing, it reduces the water permeability of the dried film.

EXTENDERS – Extenders are sometimes known as non-hiding pigments. They are added to formulations to reduce cost (saving expensive primary pigment) and, if chosen correctly, can improve the mechanical properties and application qualities of a coating.

ADDITIVES – Additives (minor components) of many different types are produced to modify specific properties of paints. They may be added to a formulation to ensure better and safer production, increase shelf life, reduce drying time and minimise sagging, for example. Two types of additives of importance in protective coatings are 'driers', which reduce the drying time of alkyd-based formulations significantly, and 'thickeners' (thixotropic agents), which are used to produce high-build formulations.

2.3 CLASSIFICATION OF COATINGS

Coatings are usually classified according to the mechanism by which they form films and then cure (or dry).

For protective coatings, there are four main mechanisms, which may act independently or in combination:

AIR DRYING : The coating cures by oxidation. Alkyds and epoxy esters, for example, form films in this way.

CHEMICAL CROSS-LINKING : The coating cures by the cross-linking of reactive organic molecules (base and hardener). Typical examples are two – component epoxies and polyurethanes. Cross-linking reactions start as soon as the two components are mixed together. Molecular weight increases and a chemically resistant film is formed.

SOLVENT EVAPORATION : The coating cures solely by evaporation of the solvent. Typical examples are vinyl and acrylated rubber coatings.

EVAPORATION OF WATER : The coating cures after the solvent (water) has evaporated. Typical examples are latex-based (polymer emulsion) coatings. Film formation involves coalescence of the non-aqueous phase.

Coatings in which the solvent is water are conventionally known as water-borne coatings. (They are also sometimes called *water-based coatings,* but current convention is used here.)

This CIRIA Report relates to general (ungalvanised) steelwork fabrication for environments normally found in building and civil engineering (*e.g.* specified in accordance with NSSS). It reflects UK legislation, its application in good practice and available materials, as at March 1997. Readers/users should note that the legislation and materials are still being developed, and will change with time.

2.4 ANTI-CORROSIVE PRIMERS

The primer is the first coat of a protective coating system. Applied to correctly prepared steel, the primer is in intimate contact with the substrate and has a direct influence on any subsequent corrosion reactions. A primer must, therefore, have good adhesion to the substrate and provide a sound foundation for subsequent coats. Primers can be pigmented with active inhibitors to prevent corrosion. The most common anti-corrosive pigments found in primers, in the UK, are zinc phosphate and metallic zinc. These pigments can be carried in a wide range of binding media.

Metallic zinc is used in the so-called 'zinc-rich primers', based most commonly on epoxy resins. In order to be capable of controlling corrosion, zinc metal primers need to contain a high percentage of zinc in the dried film. Exact percentages vary between manufacturers and the best approach is to specify that these primers comply with *BS 4652 : 1995: Specification for zinc-rich priming paints (organic media).*

The function of zinc-rich primers, and the need for high zinc content, is often misunderstood. When first applied, these coatings are porous. On exposure, the zinc starts to 'corrode', protecting the steel sacrificially. The pores then become blocked with zinc corrosion products and the film acts not only as a protective coating but also as a barrier. If the film is damaged subsequently, zinc will again protect the substrate by corroding sacrificially. For this to occur, however, the zinc must be in electrical contact with itself and the substrate.

For inhibitive primers, the inhibitor content is also of importance. The most common inhibitor now used is zinc phosphate, which can be carried in a wide variety of media such as alkyds, epoxies, and epoxy esters.

Mechanisms of inhibition are complex and are outside the scope of this Report. It is important to note, however, that all inhibitors must be sparingly soluble in water, to function effectively. Because of this sparing solubility, anti-corrosive primers should always be overcoated with intermediate coats and/or a finish. Many manufacturers produce so-called red oxide primers. These products are pigmented with red iron oxide, an inert pigment with no anti-corrosive properties. To be effective, an active pigment such as zinc phosphate must also be present.

2.5 INTERMEDIATE (OR BUILD) COATS

The purpose of intermediate (or build) coats is to increase film thickness, reduce permeability, and increase the barrier properties generally of the paint system.

Intermediate coats are often pigmented with raw materials that assist these aims specifically, such as micaceous iron oxide (MIO) or aluminium flakes. These coatings may be formulated on a variety of binder types, the most common of which are alkyds and epoxies. Lamellar (plate-like) pigments, such as MIO, decrease permeability in two ways: first, by orientation in the dried film, as discussed previously: and, secondly, by increasing film thickness. Since barrier coats do not usually contain corrosion inhibitors, they should not be applied directly to steelwork.

This CIRIA Report relates to general (ungalvanised) steelwork fabrication for environments normally found in building and civil engineering (*e.g.* specified in accordance with NSSS). It reflects UK legislation, its application in good practice and available materials, as at March 1997. Readers/users should note that the legislation and materials are still being developed, and will change with time.

2.6 FINISHES

Finishes are applied for a variety of reasons, most obviously for decorative appearance. Finishes in protective coating systems, however, may be required to provide chemical and/or abrasion resistance. Decorative finishes are pigmented to give the required colour and gloss level. In terms of corrosion protection, they add little to the overall system. When smooth and of high gloss, however, they may help water to run away from the surface, thereby reducing time of wetness. Finishes may be based on a range of resin systems. One component products are easy to overcoat and maintain. Two component products are needed, however, where good chemical resistance is required.

2.7 'COMPLIANT' COATINGS

The introduction of environmental legislation and 'compliant' coatings generally will not alter the position fundamentally, though it has already prompted the removal from the marketplace of some products (see Section 6.2).

Coating specifications will remain multi-coat in nature, with each coat still having the function outlined above. However, the higher solids content of 'compliant' materials (other than the water-borne) means that they are capable, generally, of higher film build in a single coat. Curing times can also be shorter. These two factors favour faster application times, and are a major benefit to users.

This CIRIA Report relates to general (ungalvanised) steelwork fabrication for environments normally found in building and civil engineering (*e.g.* specified in accordance with NSSS). It reflects UK legislation, its application in good practice and available materials, as at March 1997. Readers/users should note that the legislation and materials are still being developed, and will change with time.

3 Regulatory overview

3.1 INTRODUCTION

This section of the Report provides specific background information on relevant legislation relating to the protection of construction steelwork.

The following key items of UK legislation are reviewed:

The Environmental Protection Act 1990	Section 3.4
Process Guidance Note 6/23 (Coating of Metal and Plastic)	Section 3.5.3
The Geneva Protocol 1991	Section 3.6
The Chemicals (Hazard Information and Packaging for Supply) Regulations (CHIP)	Section 3.7
Control of Substances Hazardous to Health Regulations (COSSH)	Section 3.8
The Construction (Design and Management) Regulations 1994 (CDM)	Section 3.9
The Environment Act 1995	Section 3.10
The Special Waste Regulations 1996	Section 3.11.4

Legislation known to be currently under revision includes:

CHIP – CHIP97

COSHH – subject to regular revision/updating

3.2 CONSTRAINTS AND CONTROLS

Traditionally, the selection and use of coatings was based on material price and performance. For many contracts, the material price has been – and still is – regarded as the dominant factor. This approach does not reflect the true cost of coating application (see Section 8).

Regulatory aspects have now become part of selection processes. The regulatory dimension is critical even where no legislation exists. The prospect of impending legislation, reinforced by foreign precedents and practices, can act as a real constraint.

Legislative and commercial pressures are often inter-linked. The latter range from high insurance premiums to clients' preferences: clients' requirements can be as compelling as regulatory demands. Tendering for public and private procurement is now conducted on an international scale. Proof of environmental acceptability is often built into a tendering document alongside technical specifications.

This CIRIA Report relates to general (ungalvanised) steelwork fabrication for environments normally found in building and civil engineering (*e.g.* specified in accordance with NSSS). It reflects UK legislation, its application in good practice and available materials, as at March 1997. Readers/users should note that the legislation and materials are still being developed, and will change with time.

The globalisation of the coatings industry has stiffened local competition. Foreign competitors are quick to encroach on what was once considered a national preserve. The range of products offered is now international. They include a variety of protective coatings with low environmental impact. The competitive availability of such products is a commercial constraint, irrespective of regulatory demands.

General societal pressures, be they *(e.g)* from green groups or trade unions, cannot be ignored. These have now spread to industry's own trade bodies. Environmental awareness, sanctioned through life-cycle analysis, eco-auditing and environmental certification and reporting, have all entered the business ethos. Industry has responded with such umbrella initiatives as 'Responsible Care' (chemical industry) or 'Coatings Care' (paint industry).

The voluntary ban on most toxic heavy metals in coatings, or recent proposals by the UK and Dutch coatings industry to extend VOC controls to the non-industrial sector, are cases in point. Increasingly, industry is turning to self-regulation as an alternative to a prescriptive regulatory approach.

3.3 INTRODUCTION TO THE LEGISLATIVE FRAMEWORK

Legislation acts as an engine for change. At best, it also provides a harmonising principle and an even 'playing field'.

The regulatory framework in the UK is a mix of national and European-inspired legislation. The raft of controls on chemicals (such as CHIP), and implementations of EU Directives (*e.g.* COSHH and the EPA), however, are of UK origin. The EPA, in fact, was designed to anticipate EU legislation.

Figures 1 and 2 respectively indicate the range of environmental and health and safety legislation which now prevails as the context for this Report

There are few outright prohibitions affecting the use of coatings. Restrictions on products are few; most legislative pressures on coatings are indirect. They impinge on work conditions or the workplace (COSHH and CDM, for example) or on industrial processes (EPA). The EPA, however, does offer a trade-off of compliant coatings against abatement and a monitoring regime. This carrot-and-stick approach introduces commercial incentives. Such an element of choice is also present, albeit in more limited form, in the incipient *European 'Solvent Directive'.* (see Appendix 2)

The Health and Safety at Work Acts (which control emissions of noxious substances from scheduled processes) have not been included in the review as neither paint manufacture nor user currently are scheduled processes.

Controls of a different kind prevail in the string of *Dangerous Substances Directives* and their companions, the *Dangerous Preparations Directives*. The former refer to single chemicals, the latter to mixtures of chemicals (*e.g.* coatings). Their ambit is principally the classification, labelling and packaging of coatings.

The relevant UK document implementing aspects of these Directives was *CHIP 1*, in 1993, superseded by *CHIP 2* in 1994, and now supplemented and amplified by *CHIP 96* (published September 1996). These Regulations are important for coatings only insofar as they require the classification and labelling of certain dangerous substances – toxic, carcinogenic, mutagenic, for example – accompanied by suitable Risk and Safety phrases. The continual expansion of the list of dangerous substances – and the ever more stringent classification and labelling requirements – act as a disincentive to their use rather than as an outright prohibition. By classifying chromates as reprotoxic, for example, this legislation has discouraged their use.

There are many additional pieces of legislation, which affect coatings use in an indirect way. Controls are often in the form of commercial penalties. Waste is a case in point. The new *Special Waste Regulations* 1996, implementing the *EU Hazardous Waste List*, institute classification thresholds for various types of waste streams (*e.g.* toxic, flammable). Thus, by controlling the disposal of paint wastes, with its potential cost implications, the Regulations indirectly affect the use of the coatings themselves.

The flammability characteristics of solvent-based coatings are an obvious drawback, though in-house solvent reclamation units can be very cost-effective.

The principal legislation affecting the use of coatings will be outlined in greater detail in the text which follows. The impact of this legislation should be considered in conjunction with the commercial pressures which have been indicated. Their confluence serves to modify traditional performance/price considerations and leads specifiers in new directions.

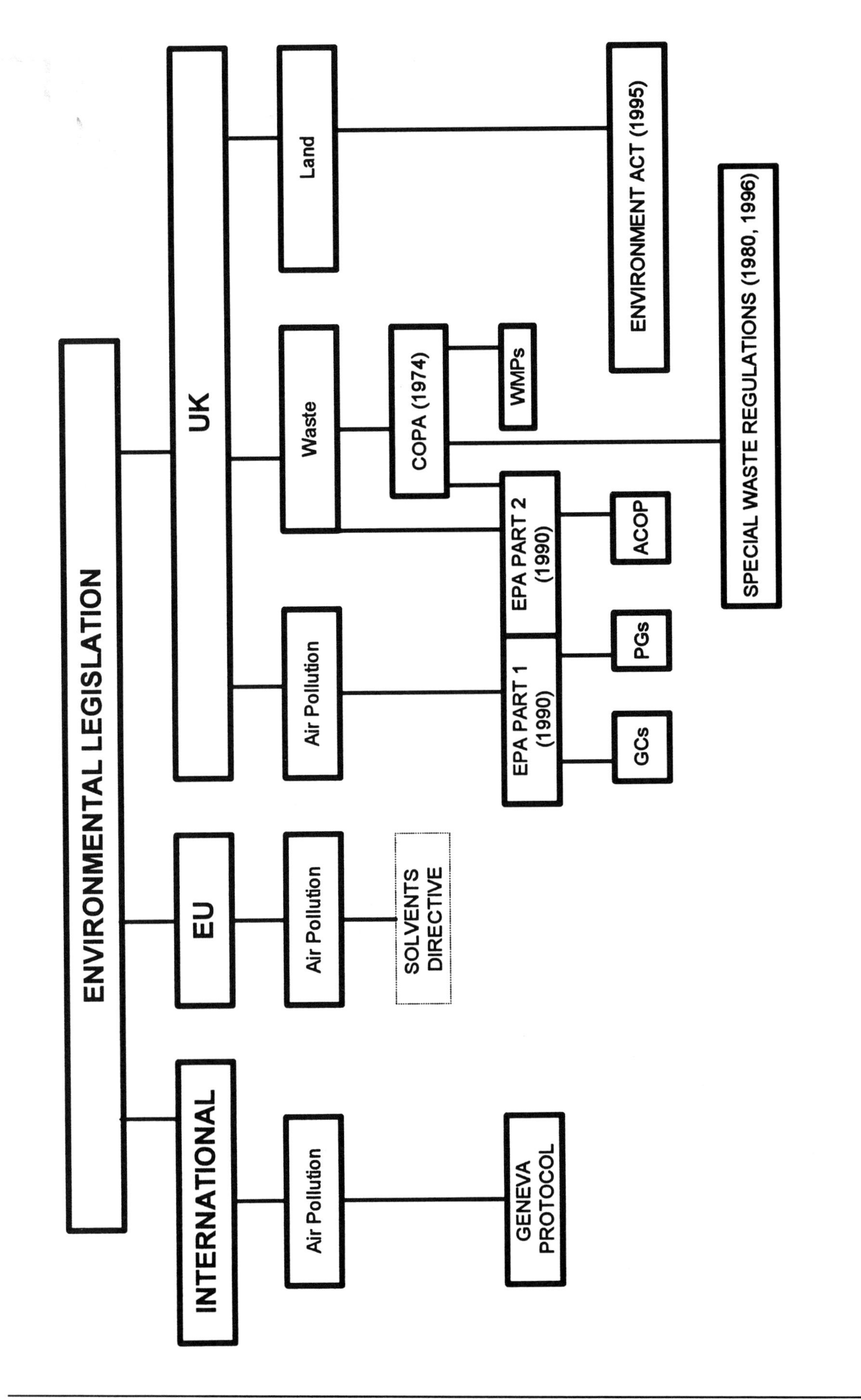

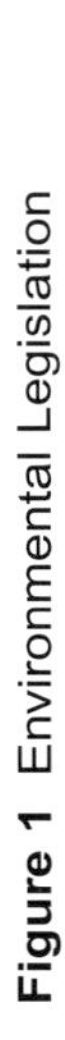
Figure 1 Environmental Legislation

This CIRIA Report relates to general (ungalvanised) steelwork fabrication for environments normally found in building and civil engineering (*e.g.* specified in accordance with NSSS). It reflects UK legislation, its application in good practice and available materials, as at March 1997. Readers/users should note that the legislation and materials are still being developed, and will change with time.

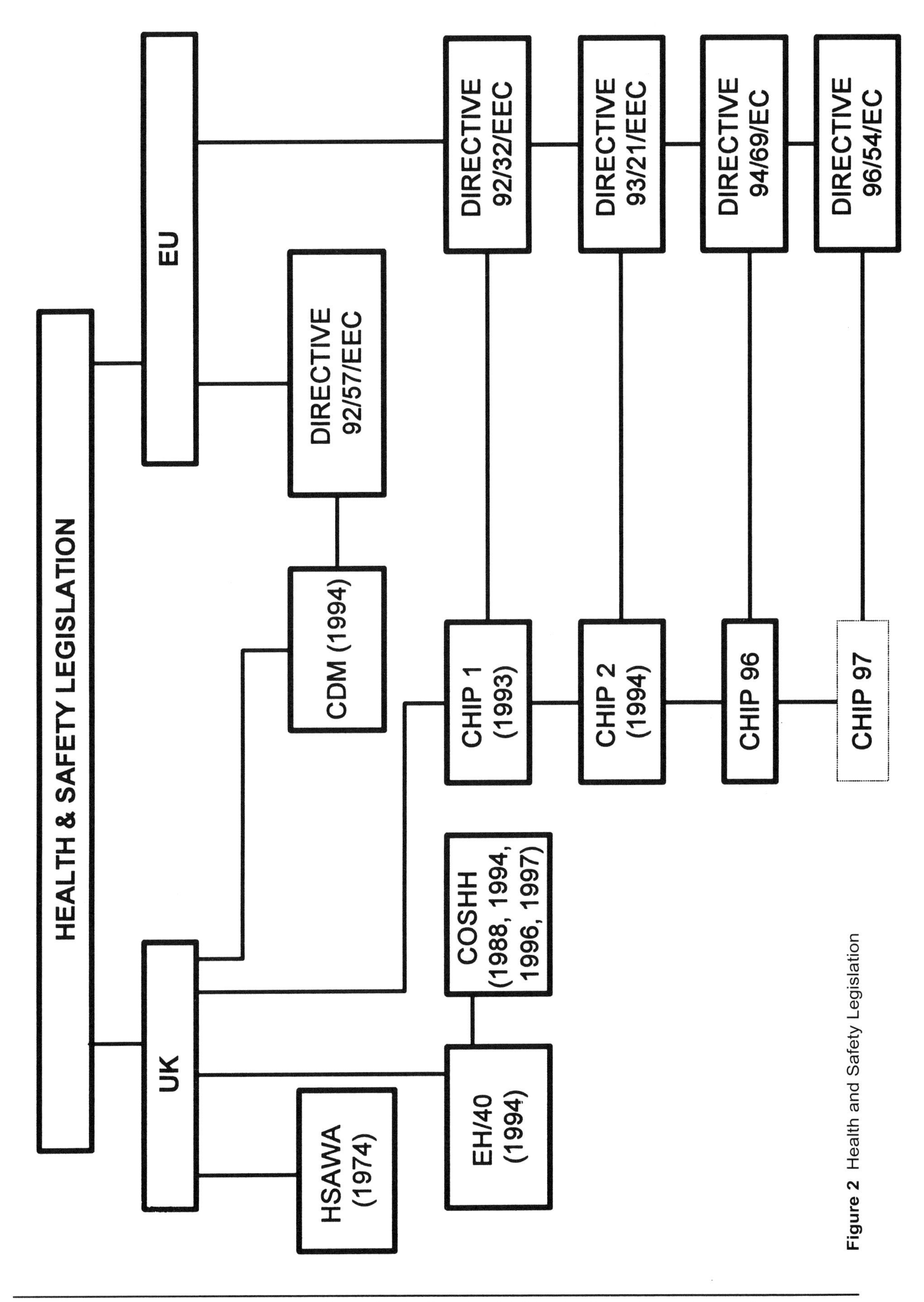

Figure 2 Health and Safety Legislation

This CIRIA Report relates to general (ungalvanised) steelwork fabrication for environments normally found in building and civil engineering (*e.g.* specified in accordance with NSSS). It reflects UK legislation, its application in good practice and available materials, as at March 1997. Readers/users should note that the legislation and materials are still being developed, and will change with time.

3.4 THE ENVIRONMENTAL PROTECTION ACT 1990

3.4.1 Scope

The most relevant documents for paints and associated processes are :

- *The Environmental Protection Act 1990, Part I* (see 3.4.4)
- *Secretary of State's Guidance Note GG1(91) – Introduction to Part I of the Act* (see 3.5.2)
- *Secretary of State's Guidance Note GG4(91) – Interpretation of Terms* (see 3.5.2)
- *Secretary of State's Process Guidance Note PG6/23* (see 3.5.3).

3.4.2 Significance

The EPA 1990 established a radical regime of pollution control extending to air, water, and land. It is a *total* approach to the environment which introduced a number of new controls and gave them punitive sanction.

The Act affects a broad spectrum of industrial activities and does not confine itself to industry alone. As a piece of omnibus legislation, it provides the framework for more detailed regulations which are contained in Statutory Instruments and are amplified in Guidance Notes.

3.4.3 Structure

The 1990 Act is in nine parts. Parts I and II deal, respectively, with *Integrated Pollution Control and Air Pollution Control* and *Waste on Land.* These two Parts form the core of the legislation. In the context of this publication, Part 1 is the relevant section.

3.4.4 The Environmental Protection Act 1990: Part I

Part 1 of the 1990 Act established two separate pollution control regimes: *Integrated Pollution Control (IPC);* and *Air Pollution Control (APC).* IPC is intended for the more potentially polluting processes and regulates releases to air, water and land; APC only regulates releases to air. Initially IPC was regulated by HM Inspectorate of Pollution (HMIP) and APC by local authorities. Under the *Environment Act 1995*, however, in England and Wales the new regulatory body for IPC is the Environmental Agency (see Section 3.10 below) while local authorities remain responsible for regulating APC. In Scotland, the Scottish Environmental Protection Agency is the regulatory body for both IPC and APC.

A wide range of industrial activities are scheduled for control ('Prescribed Processes'). Some processes are reserved exclusively for control under either IPC (Part A Process) or APC (Part B Process); others may overlap. A *Process* is the basic regulatory unit and many industrial activities will comprise a number of such processes. Paint manufacturers and paint applicators are subject to APC.

The 1990 Act makes provision for the prescription of substances for control. These substances are then prescribed in the *Environmental Protection (Prescribed Processes and Substances) Regulations 1991* (as amended) – see Section 3.10 below. Organic solvents are not prescribed as such, but volatile organic compounds (VOCs) are.

Operators of Prescribed Processes must obtain prior *Authorisation* from the relevant regulatory body before they can carry out the process. Authorisations, which are reviewed every four years, are given on the condition that operators employ *Best Available Techniques Not Entailing Excessive Cost (BATNEEC)* to prevent or minimise pollution or to render harmless any emissions. BATNEEC replaced the earlier principle of *Best Practicable Means (BPM)*.

3.4.5 Prescribed processes and substances

The most relevant documents are :

- *The Environmental Protection (Prescribed Processes and Substances) Regulations 1991 [Statutory Instrument (SI) 1991 Number 472]*
- *The Environmental Protection (Prescribed Processes and Substances) (Amendment) Regulations 1992 (SI 1992 Number 614)*
- *The Environmental Protection (Prescribed Processes and Substances) (Amendment) (No 2) Regulations 1993 (SI 1993 Number 2405)*
- *The Environmental Protection (Prescribed Processes and Substances Etc.) (Amendment) Regulations 1994 (SI 1994 Number 1271)*
- *The Environmental Protection (Prescribed Processes and Substances Etc.) (Amendment) (No 2) Regulations 1994 (SI 1994 Number 1329)*

The *Environmental Protection (Prescribed Processes and Substances) Regulations 1991,* amended 1992, specify the processes and substances for which authorisation permits are required under the *Environmental Protection Act 1990.* Chemicals, solvents and metals associated with paints and painting operations are included. Prescribed Processes are listed in Schedule 1 and Prescribed Substances released to Air, Water, and Land are listed in its Schedules 4, 5, and 6, respectively.

Coating Processes fall in Section 6.5 of Schedule 1 and the Manufacture of Coating Materials falls in Section 6.6 of Schedule 1. The full Schedules may be found in *SI 1991 Number 472.* Substantial amendments were made to the coating process definition in SI 1994 Number 1271.

Section 6.5 defines Part A Processes and Part B Processes. Part A Processes, regulated by the Environment Agency, refer to the use of coatings where there is a potential for certain listed chemicals to be emitted to water from textile or other dyeing processes or ship painting activities and fall outside the scope of this Report. Part B Processes, regulated by local authorities, refer to other uses of coatings. The relevant clauses *inter alia* from the definitions are:

PART B 'Any process (other than for the repainting or respraying of or parts of aircraft or road or railway vehicles) for the application to a substrate of, or the drying or curing after such application of, printing ink or painting or any other coating materials as, or in the course of, a manufacturing process where:

(i) the process may result in the release into the air of particulate matter or of any volatile organic compound; and

(ii) the carrying on of the process by the person concerned at the location in question is likely to involve the use in any twelve month period of:

(aa) 20 tonnes or more applied in solid form of any printing ink, paint or other coating material

(bb) 20 tonnes or more of any metal coatings which are sprayed on in molten form; or

(cc) 25 tonnes or more of organic solvents in respect of any cold set web offset printing process or any sheet fed litho printing process or, in respect of any other process, 5 tonnes or more of organic solvents'.

The amount of organic solvents used in a process shall be calculated as:

'(a) the total input of organic solvents into the process, including both solvents contained in coating materials and solvents used for cleaning or other purposes: less

(b) any organic solvents that are removed from the process for re-use or for recovery for re-use'

NOTE: These restrictions apply to the use and application of solvent-borne coatings in manufacturing processes; they do not apply to solvent-borne coatings used in an external environment (e.g. on site, or in-situ maintenance painting).

Implications of prescribed processes and substances

Industrial painting contractors, applying coatings in an enclosed environment such a fabrication/painting hall, using more than **twenty tonnes of paint in solid form (i.e powder paint)** or releasing to the atmosphere more than **five tonnes of organic solvent** in any twelve month period require an authorisation permit and are subject to the requirements of the EPA.

To put these figures in perspective, five tonnes of solvent would equate to a compliant paint consumption figure of approximately 12,500 litres (assuming a volume solids of 60% and a VOC content of 400 g/l).

This CIRIA Report relates to general (ungalvanised) steelwork fabrication for environments normally found in building and civil engineering (*e.g.* specified in accordance with NSSS). It reflects UK legislation, its application in good practice and available materials, as at March 1997. Readers/users should note that the legislation and materials are still being developed, and will change with time.

This volume of paint would be sufficient theoretically to coat 75,000 m^2 of construction steelwork at a dry film thickness of 100 micrometres (practically, 37,500 m^2 – assuming a loss factor of 50%).

In terms of typical painted tonnage, the volume equates to approximately 1,500 tonnes of steelwork *per annum*.

For a typical *non-compliant* product with a volume solids of 40% and a VOC content of 550 g/L, the allowable paint consumption would drop to just under 9,100 litres, sufficient to coat theoretically only 36,360 m^2 at 100 microns dry film thickness.

New Prescribed Processes have been subject to authorisation from 1 April 1991. They have had to comply with new plant performance standards (emission limits and operating conditions) set out in the relevant Process Guidance Note from commencement. For the majority of paint application facilities, applications for authorisations should have been made in 1992.

The 1993 Amendment to the Regulations *(SI 2405)* revised the rules of interpretation of scheduled process definitions and the responsibilities of controlling bodies. The 1994 Amendments *(SI 1271* and *SI 1329)* transferred some processes from Part A to Part B categories (and *vice versa*).

They also clarified the definition of the amount of organic solvent used in a process as: 'The total input of organic solvents into a process, including solvents contained in coating materials and solvents used for cleaning or other purposes *less* any organic solvents that are removed from the process for re-use or recovery for re-use'.

3.5 SECRETARY OF STATE'S GUIDANCE NOTES

3.5.1 Scope

Guidance to the enforcing authorities on the applications of the EPA relating to Prescribed Processes has been issued by the Secretary of State. Enforcing authorities are required to have regard for this guidance and the Secretary of State will also take it into account when considering appeals against an enforcing authority's decision. Guidance Notes have, therefore, standing in law and may be regarded as a 'highway code' for the legislation.

Two types of Guidance Note have been issued :

- General Guidance Notes
- Process Guidance Notes.

3.5.2 General guidance notes

Five General Guidance Notes (GG) have been issued so far. Of these, *GG1* and *GG4* are of most relevance to this Report. Other GGs deal with Authorisations (GG2), Applications and Registers (GG3), and Appeals (GG5).

GG1(91), issued in April 1991, is an introduction to Part I of the EPA. Guidance on the interpretation of key terms used in the legislation, notably BATNEEC. Guidance on procedural aspects of the legislation indicates interfaces with other legislation, notably Part 1 of the *Control of Pollution Act 1974* which deals with the disposal of controlled waste.

GG4(91), also issued in April 1991, provides an Interpretation of Terms used in Process Guidance Notes.

3.5.3 Process guidance notes

Introduction

In the UK, the Secretary of State for Environment has issued a series of Process Guidance Notes in support of the *Environmental Protection Act 1990, Part 1*. They are intended as a guide to local authorities on the techniques appropriate for the control of air pollution. These documents present the environmental performance standards which are expected to be achieved by new or substantially modified processes and to which existing processes are to be upgraded in terms of emission limits and process operating conditions.

The Notes specify the period for upgrading of existing processes and so are also of interest to operators of such processes. The guidance given in the Notes is based on BATNEEC assessment.

The relevant Process Guidance Note is *PG6/23 (Coating of Metal and Plastic*)

PG6/23

PG6/23 relates to processes for the coating of metal and plastic which use more than 5 tonnes of organic solvent in any twelve month period. It is directed towards the application of coatings in enclosed areas, such as spray shops and fabrication/painting halls, rather than to on-site coating, *e.g.* maintenance of structural steelwork. The Note includes emission limits and controls for organic solvents and advises on alternative approaches.

The Note applies to 'all new processes, to replacement processes, to substantial changes to existing processes and to the upgrading of existing processes to meet the standards of the Note'.

This CIRIA Report relates to general (ungalvanised) steelwork fabrication for environments normally found in building and civil engineering (*e.g.* specified in accordance with NSSS). It reflects UK legislation, its application in good practice and available materials, as at March 1997. Readers/users should note that the legislation and materials are still being developed, and will change with time.

PG6/23(92) was first issued in February 1992. This was superseded by *PG6/23(95)* and a further amended Note, *PG6/23(97)* replacing both these was published in March 1997.

PG6/23(92)

With regard to the emission of VOCs, Clause 18 of *PG6/23(92)*, which has remained substantially unchanged in the subsequent revisions, advises that:

> 'the preferred method of preventing and minimising emissions of volatile organic compounds is the reduction and eventual elimination of their use in surface coatings'.

Listed alternative approaches include:

- water-borne coatings
- high solids content coatings
- organic solvent-free liquid coatings.

Clause 19 states that the emission limits, defined in Clause 16, may be waived if coatings are applied which contain less than the maximum solvent concentration limits (usually expressed in grammes per litre) set out in Clause 19 or, for new processes, replacement processes or substantial changes to existing processes commenced before 1 April 1996, in Appendix 2. *PG6/23(95)* revised this key date to 1 April 1998.

***PG6/23(92)*, therefore, introduced the concept of 'compliant' coatings.**

NOTES: The original emission limits and the maximum VOC limits for compliant coatings given in *PG6/23(92)* were amended in *PG6/23(95)* and have been further amended and superseded by *PG6/23(97)*, see below.

The Clause numbers stated above have also been changed in *PG6/23(97)*:

PG6/23(92)	*PG6/23(97)*
Clause 16	Clause 15
Clause 18	Clause 19
Clause 19	Clause 20

PG6/23(95)

Since 1992, paint manufacturers and users have been involved in measuring, registering and preparing upgrade plans to meet the deadlines for compliance. In addition, paint manufacturers, through their trade association, the British Coatings Federation (BCF), pursued a dialogue with the UK Department of the Environment in respect of the VOC limits for 'compliant' coatings defined in the *PG6/23(92)*. BCF expressed concern that certain of the 'compliant' coatings put forward in this document were unlikely to be developed fully within the time-scale indicated. As a result of this, the Department of the Environment issued *PG6/23(95)*.

PG6/23(97)

PG6/23(97) was published in March 1997 and is the result of further consultation with the industry and the need for consistency with the proposed EU Directive (see Appendix 2 of this Report).

Clause 15 of the Note defines the emission concentration limits for release from contained sources. Those relevant to coatings for corrosion protection of construction steelwork are summarised in Table 1 below.

Table 1 Abatement emission limits for new wet painting facilities specified in Clause 15 of *PG6/23(97)*

Emission	Limit Value
Total particulate matter	50 mg/m^3
Isocyanates (total isocyanate group excluding particulate)	0.1 mg/m^3
Volatile Organic Compounds [1,2] (total carbon equivalent)	50 mg/m^3 (where the organic solvent consumption is 15 tonnes or more per annum) or 150 mg/m3 (where the organic solvent consumption is between 5 and 15 tonnes per annum)
Chlorinated Organic Solvent (total carbon equivalent)	20 mg/m^3 and 15% fugitive emissions expressed as a percentage of total solvent input to the item of equipment (where solvent use is greater than 10 tonnes per annum)

NOTES:

1 Applies to spray booths and curing ovens

2 Also applies to cleaning operations where mass emissions of VOCs exceed 1 kg in any 8 hour period

SPECIAL NOTES:

The emission limits for VOCs do not apply if compliant coatings are used (see below)

These emission limits are also not applicable where a coating process involves the use of less than 1 tonne of VOC in any 12 month period arising from the use of coatings which are non-compliant due the non-availability of compliant alternatives.

The revised figures for VOC limits for compliant coatings, which are fundamental to the rest of this Report, are given in Table 2 below.

This CIRIA Report relates to general (ungalvanised) steelwork fabrication for environments normally found in building and civil engineering (*e.g.* specified in accordance with NSSS). It reflects UK legislation, its application in good practice and available materials, as at March 1997. Readers/users should note that the legislation and materials are still being developed, and will change with time.

Table 2 VOC limits for compliant coatings in *PG6/23(97)*

Type	New processes, replacement processes and substantial changes to existing processes:	
	commenced after 1 April 1998 (Clause 20)	commenced after 1 April 1996 and before 1 April 1998 (Appendix 2)
	g/1	g/l
Etch/wash primer	780	780
Blast/weldable primer[1]	780	780
Tie coat/sealer[2]	780	780
General primer/undercoat	250	400
Topcoats	420	520

NOTES: 1 Blast primers are primers which are applied to steel immediately after it has been blasted to provide only temporary protection.

2 Tie coats/sealer are used to seal certain porous treatments of structural steel prior to painting.

Applications for authorisation made before 1 April 1998 should specify whether the operator would intend to comply with the emission limits, or use compliant coatings.

Existing processes are required to be upgraded to the standards defined in *PG6/23(97)* 'whenever the opportunity arises'. Only in exceptional circumstances should upgrading be completed later than 1 April 1998.

However, for general industrial coating and/or finishing operations using between 5 and 15 tonnes of organic solvent in any 12 month period the deadline for completion of upgrading to the emission limit requirements or VOC limits for compliant coatings has been extended to 1 April 2007, except in exceptional circumstances.

3.6 THE REDUCTION OF VOC EMISSIONS: THE GENEVA PROTOCOL 1991

The Convention on Long-Range Transboundary Pollution, adopted in 1979 and entered into force in 1983 under the United Nations Economic Commission for Europe (UNECE), lays down general principles for international co-operation on the abatement of air pollution. The Convention has been followed by a series of Protocols with specific commitments.

One of these is the *Geneva Protocol on the Control of VOC Emissions or their Transboundary Fluxes*. This commits parties to reducing their VOC emissions by 30% by 1999 from 1988 levels.

The UK, which ratified the Geneva Protocol in 1994, expects to meet this treaty obligation through the EPA and other measures.

3.7 THE CHEMICALS (HAZARD INFORMATION AND PACKAGING FOR SUPPLY) REGULATIONS

3.7.1 The CHIP regime

The regime forms the cornerstone of UK occupational legislation. So far, there have been three CHIP Regulations, in 1993, 1994 and 1996.

The revision of CHIP is continuous in line with adaptations to Technical Progress of the Dangerous Substances Directive.

3.7.2 CHIP 2

The *CHIP 2* package consists of four essential and inter-related documents.

- The Regulations, in the form of *Statutory Instrument (1994/3247)*
- Three supplementary associated publications by the Health & Safety Commission, amplifying, explaining and interpreting the Regulations.

New Regulations and an expanded Approved Supply List (the classification and labelling requirements of dangerous chemicals, together with their appropriate risk and safety phrases), *CHIP 96*, came into effect on 1 September 1996 (see 3.7.3 below)

The other two associated publications, on safety data sheets and on guidance for dealing with substances for which there is no agreed classification and labelling requirements, have not required a radical revision.

The onus is put squarely on the producers/suppliers/importers of chemicals to provide certain information. This is for the benefit of *users*, alerting them to the inherent risks of certain chemicals, and giving them the option whether or not to use them. In this way, CHIP exerts an indirect constraint on the use of coatings, insofar as they may contain certain dangerous substances.

CHIP 2 introduced new sensitising categories: sensitising by inhalation, carrying the Xn (harmful) symbol and R42 risk phrase; and sensitising by skin contact, carrying the Xi (irritant) symbol with the R43 risk phrase. *CHIP 2* also increased the number of obligatory risk and safety phrases. New phrases now cover 'toxic to reproduction' dangers.

For users of coatings, examples of relevant chemicals re-classified in *CHIP 2* include lead chromate and bisphenol A (a building block for epoxy resins). The latter has acquired the R43 risk phrase 'may cause sensitisation by skin contact'. Lead chromate had been re-classified as 'toxic for reproduction' with a new danger indication – T (toxic) instead of Xn – as well as new R61 risk phrase ('may cause harm to the unborn child') and additional safety phrases. In paints, the concentration refers to the percentage by weight of the metallic element calculated with reference to the total weight of the preparation.

This CIRIA Report relates to general (ungalvanised) steelwork fabrication for environments normally found in building and civil engineering (*e.g.* specified in accordance with NSSS). It reflects UK legislation, its application in good practice and available materials, as at March 1997. Readers/users should note that the legislation and materials are still being developed, and will change with time.

3.7.3 CHIP 96

CHIP 96, the *Chemicals (Hazard Information and Packaging for Supply (Amended) Regulations 1996,* came into force on 1 September 1996, implementing the 21st Adaptation to the Dangerous Substances Directive. It provides new Regulations and a new 1996 *Approved Supply List* which contains both modified and new entries. Substances classified as dangerous for the first time include certain paint solvents, including white spirit. Three white spirit types are regarded as Category 2 carcinogens (R45), but not at the concentration thresholds typically used in paints. Zinc chromate, considered as a Category 1 carcinogen, has a new labelling phrase.

3.7.4 Draft CHIP 97

The 22nd Adaptation to the Dangerous Substances Directive (Directive 96/54/EC, OJ, L248 30 September 1996) was published at the end of September 1996. The Draft CHIP97 Consultative Document, has been published and includes draft Regulations, draft Approved Supply List and the revised Approved Classification and Labelling Guide. CHIP97 will be the second amendment to the CHIP2 Regulations of 1994.

3.8 CONTROL OF SUBSTANCES HAZARDOUS TO HEALTH REGULATIONS (COSHH)

3.8.1 The Regulations

The *Control of Substances Hazardous to Health Regulations* (COSHH) provide a framework to protect people in the workplace against health risks from hazardous substances. COSHH lays down a step-by-step approach to the necessary precautions to be taken. It sets out essential measures which employers and employees must adopt. Failure to comply with COSHH can put people at risk and is an offence under the *Health and Safety at Work Act 1974* (HSWA).

COSHH applies to virtually all substances hazardous to health. This includes paints, which are classed as '*hazardous substances*'.

Excepted are asbestos and lead (which have their own regulations), and substances which are hazardous only because they are radioactive, asphyxiants, at high pressure, at extreme temperatures, or have explosive or flammable properties.

Consolidated COSHH Regulations were published in 1994 and came into force on 16 January 1995.

This CIRIA Report relates to general (ungalvanised) steelwork fabrication for environments normally found in building and civil engineering (*e.g.* specified in accordance with NSSS). It reflects UK legislation, its application in good practice and available materials, as at March 1997. Readers/users should note that the legislation and materials are still being developed, and will change with time.

The 1994 COSHH Regulations incorporate:

- the original 1988 COSHH Regulations *(SI Number 1657)* [both the 1991 and *1992 COSHH (Amendment) Regulations (SI Numbers 2431* and *2382*, respectively].
- a revised definition of 'a substance hazardous to health' in line with the *Chemicals (Hazard Information and Packaging) Regulations 1993* (CHIP). The extension of COSHH to pipelines and associated work activities.

The Regulations are supported by a number of Approved Codes of Practice (ACOPs). Also relevant is: CIRIA Report 125: *A guide to the control of substances hazardous to health in construction*, which provides guidance on these Regulations.

Since 1994, two amendments have been published:

The Control of Substances Hazardous to Health (Amendment) Regulations 1996
The Control of Substances Hazardous to Health (Amendment) Regulations 1997

3.8.2 Hazards and risks

The words '*hazard*' and '*risk*' have special meaning within the COSHH Regulations. The *hazard* presented by a substance is its potential to cause harm. The *risk* from a substance is the likelihood that it will cause harm in the actual circumstances of use.

Risk will depend upon :

- the hazard presented by the substance
- how it is used
- how exposure to it is controlled
- to how much of the substance one is exposed and for how long
- whether one is sensitive to it.

In reality the situation is more complex. There may be a substantial risk from a substance which is not especially hazardous, if exposure is prolonged. Conversely, the risk of being harmed by the most hazardous substances may be very small, if proper precautions are taken.

3.8.3 Complying with COSHH

Complying with COSHH involves :

- assessing the risks to health which may arise from any work
- deciding what precautions need to be taken (work should not be carried out which could expose staff to hazardous substances unless the risks have first been assessed and the necessary precautions taken)
- preventing or controlling the risks
- ensuring that control measures are used and maintained properly

- ensuring that clearly defined safety procedures are followed
- monitoring exposure of workers to hazardous substances (health surveillance)
- informing, instructing and training staff about the risks and the precautions needed.

3.8.4 COSHH assessment

Step-by-step compliance with COSHH is achieved by assessments. Hazards are first defined and risks are then assessed. Where significant risks are found, then action is needed to reduce or eliminate them.

The responsibility for COSHH assessments rests ultimately with the employer, though other parties may carry out the actual assessments on behalf of the employer. The assessor (often a professional consultant) needs to :

- have access to the COSHH Regulations and Approved Codes of Practice
- understand these documents fully
- have the ability and the authority to assemble all the necessary information and then make informed decisions about the risks involved and the precautions needed.

3.8.5 The relevance of COSHH to coatings and their application

In terms of COSHH, substances hazardous to health include, *inter alia*, chemicals with occupational exposure limits (OELs). OELs are defined by the Health and Safety Executive in *EH40/94 (Occupational Exposure Limits 1994)*. Organic solvents and some two-component resin systems used as binders in coatings (*e.g.* epoxies and polyurethanes) fall within this category. Particular care is needed, therefore, in the storage and use of solvent-borne coatings; other coating types may also be hazardous to health, if handled unwisely.

The employer of paint applicators has a responsibility to supply sufficient information and instruction on :

- control measures (their purpose and their use)
- personal protective equipment and clothing
- emergency procedures
- monitoring/health surveillance (and the results therefrom).

Whilst it is the responsibility of the contractor to assess the risk, personnel carrying out the work (*i.e.* the painters) should also make individual assessments. In this respect, useful information may be found in the Technical Data Sheets and the Health and Safety Data Sheets issued by paint manufacturers.

Local offices of the Health and Safety Executive and the local Environmental Health Officer are always willing to assist. Their advice should always be sought whenever clarification is needed.

3.9 THE CONSTRUCTION (DESIGN AND MANAGEMENT) REGULATIONS 1994

3.9.1 Introduction

The *Construction (Design and Management) Regulations 1994 (SI Number 3140)* also known as the CDM Regulations, came into force on 31 March 1995. The Regulations, which implemented *EC Council Directive 92/57/EEC*, aim to provide for minimum health and safety standards at temporary or mobile construction sites. They also put into place provisions which are the result of an extensive review of existing construction-related legislation by the Health and Safety Commission (HSC), through its Construction Industry Advisory Committee (CONIAC).

The CDM Regulations create a new safety planning role – **the Planning Supervisor**. It is his/her function to ensure that safety considerations are managed as part of the design process. The Planning Supervisor has overall responsibility for co-ordinating the health and safety aspects of the design and planning phase, and responsibility for the early stages of the *Health and Safety Plan* and the *Health and Safety File*.

In a typical project involving construction steelwork, the Planning Supervisor, who has responsibility throughout the entire project, can be appointed from the project management, engineering design or the architect's team. Independent appointments are also made.

The Regulations also place specific duties on another **four** key parties:

1. **The Client** is responsible for providing information in his possession relevant to Health and Safety, *e.g.* site and operational conditions. He should also be satisfied that only competent people are appointed as planning supervisor and principal contractor. The Client is required to ensure that sufficient resources, including time, have been or will be allocated to enable the project to be carried out safely.
2. **The Designer** should ensure that structures are designed to avoid or minimise risks to health and safety while they are being built or maintained. Where risks cannot be avoided, adequate information has to be provided. Design is not limited to drawings.

The Designer is responsible for the preparation of Specifications, or approval of these by others, including paint specifications.

During the design process, the Designer must :

- Identify hazards that might result from the design (*e.g.* a coating which will be difficult to maintain).
- Avoid, reduce or control the hazard, if possible (*e.g.* change the method of application).

This CIRIA Report relates to general (ungalvanised) steelwork fabrication for environments normally found in building and civil engineering (*e.g.* specified in accordance with NSSS). It reflects UK legislation, its application in good practice and available materials, as at March 1997. Readers/users should note that the legislation and materials are still being developed, and will change with time.

- Carry out a risk assessment of the hazard remaining (*i.e.* the severity and the probability).
- Suggest the preferred preventative and protective measures (*e.g.* personal protection equipment).
- Identify hazardous materials or substances to be used (*e.g.* isocyanates).
- Designers should record their consideration of the above matters formally and pass that information on to others. A Hazard Identification and Risk Assessment sheet is often used for this purpose.

3. **The Principal Contractor** is generally the **Main Contractor**. He has overall responsibility for the management of site operations. The Main Contractor should take account of health and safety issues when preparing and presenting tenders or similar documents. In addition, he should also develop the *Health and Safety Plan* and co-ordinate the activities of all contractors to ensure that they comply with health and safety legislation. Main Contractors have a duty to check on the provision of information and training for employees and for consulting with employees on health and safety.
4. **Sub-Contractors** should co-operate with the Main Contractor and provide relevant information on the health and safety risks created by their work and how they will be controlled. Sub-Contractors also have duties for the provision of other information to the Main Contractor and to employees.

3.9.2 Scope and purpose of the Regulations

CDM Regulations apply to construction work which is *notifiable* – whole or part contracts in which the construction phase will occupy more than thirty days or will involve more than five hundred man days of work. CDM Regulations also apply, however, to *non-notifiable* work involving five people or more on site at any one time. Design work, of any length and any team size, is also included.

The main purpose of the Regulations is to establish a safety management network at all stages of a project.

The CDM Regulations impose duties and involve all parties who can contribute to Health and Safety on a construction project. In particular, the Client and the Designer have designated duties.

The Client is not usually the Principal Contractor, unless he has in-house maintenance or a construction group with a site management function.

3.9.3 Health and Safety Plan and Health and Safety File

The Planning Supervisor is required to prepare a *Health and Safety Plan* and *a Health and Safety File*. These documents are then developed and kept up to date by the Principal Contractor. The *Health and Safety Plan* must, amongst other things, provide information to assist contractors to plan construction and protective coating procedures.

The *Health and Safety Plan* provides the health and safety focus for the construction phase of a project. At the pre-tender stage, relevant health and safety information must be available for contractors tendering or making arrangements to carry out or manage construction work. The responsibility for this rests with the Planning Supervisor.

The *Health and Safety File* is a record of information for the client/end user. It informs those who might be responsible for the structure in the future of the risks that have to be managed during maintenance, repair or renovation.

At the end of the project, the *Health and Safety File* is handed over to the client. The client must then make the document available to those who will work on any future maintenance, modification, or demolition of the structure.

3.9.4 Protective coating systems

Within the context of the CDM Regulations, solvent-borne protective coatings are considered as 'Hazardous Substances'.

The Designer, as one of their key duties, has a responsibility (so far as is reasonably practicable) to design to avoid risks. The achievement of this will involve consideration not only of the construction phase but also how the completed structure will be maintained, repaired and eventually removed.

The guidance given by the Construction Industry Advisory Committee in *Designing for Health and Safety in Construction* (A guide for designers on the CDM Regulations 1994, prepared in consultation with the Health and Safety Executive) includes the following advice :

- Specify water-borne coatings, wherever practicable, as they are less hazardous (safer) generally than organic solvent-borne coatings.
- Specify shop application (where conditions can be controlled more easily).
- Do not specify organic solvent-borne coatings for use in confined or difficult to ventilate areas.

3.10 THE ENVIRONMENT ACT 1995

The Act, which received Royal Assent on 19 July 1995, established for England and Wales, the Environment Agency (the 'Agency') and for Scotland, the Scottish Environment Protection Agency (SEPA).

As of April 1 1996, it transfers the functions of the NRA, which, along with those of Her Majesty's Inspectorate of Pollution (HMIP) and the Waste Regulation Authorities (WRAs), were taken over by the Environment Agency in England and Wales. Its equivalent functions in Scotland (including those of the river purification boards) were taken over by the Scottish Environmental Protection Agency (SEPA), and in Northern Ireland by the Environment and Heritage Service.

The new Agencies have a duty to exercise their pollution control powers to:

- prevent, minimise, and/or remedy environmental pollution,
- monitor environmental pollution,
- follow developments in environmental technology,
- assess the effects of environmental pollution,
- report on pollution control options.

The Agencies also have the power to provide advice or assistance to the public, and to establish charging schemes.

Part II of *The Environment Act 1995* defines 'contaminated land' and makes provisions for its regulation. EPA 1990, added a new Part IIA with sixteen sections. The definitions and procedures for statutory nuisances contained in Part III of the EPA are thus replaced.

Two types of contaminated land are distinguished :

- special sites under the appropriate Agency
- all other contaminated land and under local authority control.

Remediation notices for enforcing the clean-up of contaminated sites can specify procedures and the time period. The 'appropriate person' on whom such a notice can be served is the person 'who caused or knowingly permitted' the contamination, or if no such person can be found, the owner or current occupier of the contaminated site. A third party may be required to clean up a site due to contamination by migration from an adjoining site.

Compliance will be spelled out in more detail in Guidance Notes and Regulations, which are expected to be published in 1997.

3.11 WASTE MANAGEMENT

3.11.1 General

All waste has the potential to affect the environment adversely, by contaminating air, land, and water. The term 'waste' can cover anything from household waste to hazardous waste from industrial processes.

The *Control of Pollution Act 1974* (COPA), Part I, was the primary legislation controlling waste collection and disposal in the UK. This Act has now been replaced by Part II of the EPA.

In essence, Part II of the EPA :

- establishes Her Majesty's Inspectorate of Pollution (HMIP) and the Waste Regulatory Authority as the regulatory bodies
- provides for waste management licensing to keep, treat or dispose of controlled waste
- imposes a Duty of Care on importers, producers, keepers, treaters or disposers of controlled waste.

Hazardous (or toxic) waste – known as 'special waste' – is subject to additional stricter controls under the *Control of Pollution (Special Waste) Regulations 1980 (SI 1709), The Special Waste Regulations 1996 (SI 972)* and *The Special Waste (Amendment) Regulations 1996 (SI 2019).*

'Special Waste' Regulations establish a prior notification and consignment note system for transfers of special waste. (Materials dangerous to life, or materials with a flash point below 21°C, are classified as special waste). Special waste may include, therefore, waste from painting operations, chlorinated solvents, and heavy metals.

The definition of hazardous waste is currently under review, in the light of the amendments and expansions which have been issued by the EC to its Framework Directive on Waste *(75/442/EEC).*

3.11.2 Waste management papers

The Department of Environment has prepared a series of guidance papers to provide technical guidance on all aspects of waste management, including legislative requirements and advice on the treatment and safe disposal of waste. These papers were first issued in support of the *Control of Pollution Act 1974.* They are now being updated to take account of the EPA 1990 and other technical developments.

There are currently 28 papers in the series on all aspects of waste management *(WMP 1* to *WMP 28).*

In the context of this Report, two WMPs may be of potential interest to applicators of protective coatings. These are:

WMP 14 (1977) Solvent Wastes (excluding Halogenated Hydrocarbons) – a Technical Memorandum on Reclamation and Disposal

WMP 15 (1978) Halogenated Organic Wastes – a Technical Memorandum on Arisings, Treatment and Disposal

These papers deal *inter alia* with the reclamation and safe disposal of organic and halogenated organic solvents, respectively. Coatings may contain organic solvents and halogenated organic solvents may be used to strip coatings.

This CIRIA Report relates to general (ungalvanised) steelwork fabrication for environments normally found in building and civil engineering (*e.g.* specified in accordance with NSSS). It reflects UK legislation, its application in good practice and available materials, as at March 1997. Readers/users should note that the legislation and materials are still being developed, and will change with time.

3.11.3 Code of Practice on the Duty of Care Aspect of Waste Management

The Department of Environment has also prepared and issued (HMSO, March 1996) a *Code of Practice on the Duty of Care Aspect of Waste Management* in support of the EPA. The purpose of the code is to set out practical guidance for waste holders subject to the duty of care in respect of *controlled waste* (Section 34 of the EPA 1990). The code is divided into :

- step by step advice on following the duty
- a summary checklist
- annexes.

The annexes describe :

- the law on the duty of care
- responsibilities under the duty
- regulations on keeping records
- an outline of other legal requirements
- a glossary.

The most important point to note is that the onus is on the Waste Maker to ensure that disposal is being carried out correctly, even if this is being carried out by a third party.

3.11.4 The Special Waste Regulations 1996

The Regulations of 1 April 1996 *(SI 972)* came into force on 1 September 1996, are a composite document, implementing, on the one hand, the *EC Directive 91/689/EEC* on hazardous waste and the hazardous waste list *(94/904/EC)* and amending, on the other, UK legislation (*e.g.* the *Waste Management Licensing Regulations 1994)*. They modify substantially (but do not supersede completely) the *Control of Pollution (Special Waste) Regulations 1980*. Their essential feature is that they deal only with the control of **hazardous – i.e. special** – waste.

The main provisions of the 1996 Regulations are:

- A new definition of special waste, incorporating the EC Hazardous Waste List. To be 'special', waste must have the hazardous properties set out in the 1991 Directive. Some wastes, not on the EC List, are carried over from the 1980 Regulations.
- A revised consignment note carrying a full description of the waste and its hazards. This note will travel with the waste to its final destination. The consignment note replaces the duty of care note in the case of special waste.
- A new system to allow for repetitive movements of waste, allowing carriers to pre-notify door-to-door collection rounds. These provisions, will replace the existing 'season ticket'.
- A prohibition on the mixing of different categories of special waste, and the mixing of special and non-special waste, without prior authorisation.
- Charges in line with the 'polluter pays' principle.

In particular, the Regulations contain a description of hazardous wastes from different industrial processes. Wastes from the manufacture or use of paint are divided into waste paints and sludges from paint, further sub-divided into those containing or free from halogenated solvents. Hazardous properties (*e.g.* toxic, flammable, oxidising, corrosive) are defined in another part of the schedule.

A Guidance Note *(DOE Circular 6/96)*, for use with the Regulations, was published on 13 June 1996.

An amendment to these Regulations, *The Special Waste (Amendment) Regulations 1996 (SI 2019)* was issued on 1 August 1996. The Amendment clarified and update *inter alia* the cross references to the Approved Supply List to ensure compatibility with CHIP96

This CIRIA Report relates to general (ungalvanised) steelwork fabrication for environments normally found in building and civil engineering (*e.g.* specified in accordance with NSSS). It reflects UK legislation, its application in good practice and available materials, as at March 1997. Readers/users should note that the legislation and materials are still being developed, and will change with time.

4 Controlling VOC emissions from coatings

4.1 THE OPTIONS

The EPA: Part I specifies two broad strategies for reducing VOC emissions. These can be considered as 'front end controls' and 'end of pipe controls'.

Front end controls are those which reduce solvent emissions at source. Examples include :

- the use of compliant coatings
- improving transfer efficiency by the use of electrostatic paint spraying equipment (where practicable)
- containment – applying coatings at works in preference to site
- improved housekeeping/solvent management.

End of pipe controls capture or destroy VOCs from waste streams after they have been generated. Abatement methods include:

- incineration
- condensation
- absorption
- adsorption
- bio-digestion.

Incineration is the most common abatement method, though it produces carbon dioxide and nitrogen oxides which are themselves pollutants. The procedure is not an option when coatings are applied on site.

4.2 THE CASE FOR COMPLIANT COATINGS

4.2.1 What is a compliant coating?

In simple and strict terms, a compliant coating is one that meets the requirements of the EPA, as defined in *PG 6/23* (see Section 1.2). Only this PG gives a definition of compliance in terms of the permitted VOC levels. In the wider context, a compliant coating would also help industry meet its wider needs and concern, to:

- facilitate compliance with COSHH and CDM legislation
- recognise the needs of current and emerging fabrication practice
- provide at least equivalent performance to existing materials with no significant cost penalty.

This CIRIA Report relates to general (ungalvanised) steelwork fabrication for environments normally found in building and civil engineering (*e.g.* specified in accordance with NSSS). It reflects UK legislation, its application in good practice and available materials, as at March 1997. Readers/users should note that the legislation and materials are still being developed, and will change with time.

These wider aims are less easily defined, but are no less important.

Reductions in VOCs should make it easier to comply with various health and safety legislation (see also Section 3.8). Similarly this reduction will tend to make it easier to integrate coating operations with other shop operations.

Meeting the requirements of *PG6/23* does not imply that a material is safe or avoid the need to comply with any other legislation, but only that the VOC limits in *PG6/23* are met.

The relevant legislative requirements of the EPA and related legislation has been discussed in Sections 3 and 4. Initially, this legislation did not define the concept of a compliant coating; the original legislation treated the construction industry in the same manner as other sectors of manufacturing. The steel construction industry – particularly the coating manufacturers – quickly realised the practical and cost implications of the original legislation. Change would have required massive capital expenditure on abatement equipment by fabricators. In effect, complying with the requirements would have been impractical by this method.

The industry persuaded Government to adopt a wider view that encompassed the needs of construction, at the same time meeting the main thrust of the EPA to reduce solvent emissions by the end of the century. The concept of a 'compliant' coating came about by widening the definitions within the Act, through the limits given in *PG6/23*. The coating industry argued that it was much better to reduce the VOC content of the materials used, rather than recycle the solvents from existing materials. The actual limits on solvent contents, for a coating to be regarded as VOC compliant, have changed over time and the values could be changed in the future at relatively short notice.

The adoption of the concept of compliance was a success for the coating industry. However, it would be wrong to conclude that the coating industry suddenly became converted to the environmental cause by the EPA. They had already recognised the need for change, but saw the situation as an opportunity to encourage change in a manner to which it was already committed.

4.2.2 Traditional specifications

The steel construction industry has never had a standard specification, or a single guidance document, aimed specifically at general steel fabrication. There are a wide variety of specifications available to the industry. The most commonly encountered are:

- *BS 5493*. A thorough and detailed document that can be used in a wide range of situations. However, it has a number of drawbacks: it is not specifically aimed at the construction industry, is complex and to a large extent is now out of date.
- *Department of Transport Specification for Highway Works*. Primarily aimed at bridge fabrication. It therefore addresses particular concerns related to these

fabrications. As such, it is largely inappropriate for more general steel fabrication.

- *British Steel technical literature.* British Steel have produced a number of simplified guidance documents aimed at the general steel fabrication industry. These documents are easy to understand and to use; they have been widely adopted in the industry. However, they are not always regarded as being totally independent as British Steel have – quite properly – an interest in promoting the use of steel.

These specifications cover a wide range of different materials, with different performance characteristics in different environments. The temptation therefore exists for the specifier to 'design' the best system for a given situation, resulting in a number of different specifications even for a single project. The fabricator is therefore often faced with a range of specifications on different projects for essentially the same service conditions.

4.2.3 Benefits of compliant coatings

One potential benefits of the EPA is that the *range* of *PG6/23* compliant materials available is considerably reduced. In effect, the choice for primers and build coats (*i.e.* not finishes) for general construction is between the water-borne or the high-solids epoxy materials.

The adoption of *PG6/23* – compliant coatings as standard materials in the construction industry, would result in rationalisation of specifications and remove an unnecessary area of contention and potential dispute. It could also improve productivity, without a sacrifice in performance.

The two most important factors in determining the performance of a given type of coating (that also significantly influence costs) are the quality of surface preparation and the total applied thickness. Other factors influencing performance are identified in Section 9.3.

The quality of surface preparation is currently established as blast cleaning to *BS7079 Part A1, grade Sa 2½*. This will not change with the introduction of compliant materials.

Most modern specifications for coatings rely on a film thickness in the range 100 to 250 microns. As conventional materials generally have only limited film building characteristics, it is necessary to apply multiple coats of paint to achieve a given thickness. This has a direct effect on application time and therefore on costs.

Compliant high-solids coatings will achieve comparable thicknesses in fewer coats. This results from the much improved film-building characteristics.

This CIRIA Report relates to general (ungalvanised) steelwork fabrication for environments normally found in building and civil engineering (*e.g.* specified in accordance with NSSS). It reflects UK legislation, its application in good practice and available materials, as at March 1997. Readers/users should note that the legislation and materials are still being developed, and will change with time.

It is this ability to achieve higher thickness in fewer coats that is one of the advantages of compliant coatings. This is not true of the water-borne materials; the currently available versions have similar film building characteristics to conventional materials.

4.3 TODAY'S PRODUCTS

A review of manufacturers' technical data sheets produced during the 1990s reflects the commitment to producing and launching high-solids materials and to providing industry with the products they need to comply with *PG 6/23*. It reveals an increasing number of products that meet the VOC compliance requirements of 1996 and, more recently, those for 1998.

The manufacturers whose major market is the construction industry now have a full range of materials that are 1996-compliant; most are close to having their full range as 1998-compliant. For primers and build coats these product ranges are based on either high-solids epoxy materials or water-borne epoxy or acrylic resins.

These products can usually be easily identified in the product manuals of the manufacturers, they nearly all contain titles that indicate the fact they are compliant in their descriptive labels. Other than these labels, there is little at first sight to suggest that the products are any different to previous product ranges. The data sheets will contain a description of the product and these will be the same as the solvented predecessor, for example:

Primers:	Epoxy Zinc Rich
	Epoxy Zinc Phosphate
Build Coats:	Epoxy MIO

Closer inspection reveals that all these data sheets state that the material meets either the 1996 or 1998 requirements of the *PG 6/23* (see Section 3.5.3). It will only quote one date, but if it is 1998 then it automatically meets the less stringent 1996 requirements.

The data sheet will also normally quote both the volume solids and VOC content of the material. It is thus easy for the specifier, or user, to establish whether a product is compliant or not by checking the data sheet against the requirements of *PG 6/23* for the particular product type.

If the information cited above is not quoted on the data sheets then it is reasonable to assume that the material is non-compliant.

5 Current materials

In the previous Chapter, it was argued that the adoption of 'compliant specifications' should be viewed in a positive manner. This Chapter of the Report reviews the position at August 1996, with regard to the materials currently being used in industry; it also highlights those materials appropriate to general fabrication, and why other materials are inappropriate.

5.1 AVAILABLE COMPLIANT MATERIALS

The range of compliant materials available to the end user can appear very wide. A cursory review of a typical manufacturer's product manual might suggest that there is a wide range of different product types that are marketed as compliant. A selective list of generic types is given in Table 3. Despite this range of materials, most are *inappropriate* for routine use in the construction industry.

5.1.1 Single-coat high-build materials

None of the single-coat high-build (500 to 1000 microns) materials are used in the general steel construction industry. These products are more commonly encountered in the offshore, petrochemical and pipeline industries and, increasingly, in bridge construction and other large civil engineering projects, particularly for maintenance applications. In these cases, the additional capital costs can be more than offset by the increased performance and reduced maintenance over the life of the structure.

5.1.2 Powder coatings

The necessary specialist equipment, for both application and curing of the coating, is not readily available to fabricators, but remains the preserve of a few specialist applicators. The use of powders is limited by the size of the section to be coated: for structural steel sections, this is undoubtedly a limiting factor. However, powder coatings are encountered in the construction industry – as architectural finishes for cladding materials such as aluminium, pre-galvanised steel strip. These are usually based on either polyester or polyvinylidenefluoride (PVF_2) resins. They are primarily used for surface appearance and are not currently used on structural steel.

5.1.3 Water-borne materials

The water-borne materials – either acrylic or epoxy – are in many ways ideal choices for use in steel construction. Containing little or no volatile organic solvent, they easily comply with the EPA. They relieve the fabricator/applicator of many of the requirements of COSHH, CDM and other related Health and Safety legislation relating to solvents. However, they are not *inherently* safe, as they are still based on epoxy materials that are potentially hazardous. They have found favour in other parts of the world, most notably certain States in the USA where environmental legislation is particularly strict. In the UK, these materials have become increasingly available during the 1990s. Water-borne materials are discussed in more detail in Sections 6.1.2 – 6.1.4.

5.1.4 Polysiloxanes

At the time of compiling this Report, a range of new products were starting to become available in the UK. These are the polysiloxane copolymers. These have not been widely used in the UK. Their advantages are that they are compliant, typically 150 g/litre VOC, and provide a high-build film with excellent cosmetic properties. If used in conjunction with a zinc silicate primer, they provide a completely inorganic system of protection, which also has the advantage of being applied in only two coats. At present, they are expensive for general construction and are only available from a single supplier, but may become more widely available.

5.1.5 Pre-fabrication primers

This is a special category of materials that are generally based on solvented epoxies or zinc silicates, with very low solids contents and film-build properties. These will remain essentially unchanged by the current legislation, as the VOC limit for these materials is presently set at 780 (g/litre). Also available in this category are polyvinyl butyral (PVB) etch primers either as single or two pack materials. However, water-borne materials are available and it is likely that, at some time in the future, stricter control of VOCs will see an increase in use these materials.

5.1.6 High solids epoxies

These are currently the most widely used in construction. These materials are likely to dominate the market for the foreseeable future. Where appearance is important they maybe overcoated to improve resistance to chalking, caused by ultra -violet light. The most common finish that is compliant, are the acrylic urethanes. However, other finishes may also be available. High solids epoxies are discussed in more detail in Section 6.1.1.

5.2 INORGANIC COATINGS

Inorganic coatings are not based on the type of organic binder systems that are commonly encountered in the construction industry. The range of products available in this form is much more limited. Traditionally, the only inorganic coatings likely to be encountered were the zinc silicate primers. These materials are not commonly used in general UK construction but are available both as water-borne and solvented materials.

This CIRIA Report relates to general (ungalvanised) steelwork fabrication for environments normally found in building and civil engineering (*e.g.* specified in accordance with NSSS). It reflects UK legislation, its application in good practice and available materials, as at March 1997. Readers/users should note that the legislation and materials are still being developed, and will change with time.

5.3 AVAILABILITY OF COATINGS THAT ARE NON-COMPLIANT

The range of materials that can be produced as compliant and are appropriate to general construction is limited. It is probable that some currently available materials that **cannot** be produced as compliant materials will gradually be removed from product ranges.

These include those materials based on chlorinated rubbers, conventionally-solvented acrylated rubbers and vinyls. The situation regarding alkyds is less clear cut: the use of alkyd primers is likely to be phased out unless suitable high solids materials can be developed to meet the compliance requirements. The well proven and established alkyd finish coats, however, will remain in the foreseeable future.

Table 3 Examples of compliant coatings

(N.B This Table is indicative, and not exhaustive with only a sample of advantages and disadvantages included)

Generic Material	Typical Thickness	Typical VOC Content	Applications	Advantages	Disadvantages	Comments
SINGLE COAT HIGH-BUILD MATERIALS						
Solvent-free polyurethanes (refer to Section 5.1.1)	> 1mm	Zero	Pipelines, offshore and petrochemical, bridge maintenance	Solvent free and high film build in a single coat.	Requires specialist application equipment. Expensive first cost	Unlikely to be used unless cost can be justified by decreasing life cycle costs
Fusion bonded epoxy (refer to Section 5.1.1)	> 1mm	Zero	Pipelines, offshore and petrochemical, bridge maintenance	Solvent free and high film build in a single coat	Requires specialist application equipment. Expensive first cost	Unlikely to be used unless cost can be justified by decreasing life cycle costs.
Modified low solvent epoxies (conventional) (refer to Section 5.1.1)	400-1000 microns	1998 compliant <150g/l	Pipelines, offshore and petrochemical, bridge maintenance	Very low VOC and high film build in a single coat	Expensive first cost	Unlikely to be used unless cost can be justified by decreasing life cycle costs.
Glass flake epoxies (refer to Section 5.1.1)	400-1000 microns	1998 compliant <150g/l	Pipelines, offshore and petrochemical, bridge maintenance	Very low VOC and high film build in a single coat	Expensive first cost	Unlikely to be used unless cost can be justified by decreasing life cycle costs.
Glass flake vinyl esters (refer to Section 5.1.1)	400-1000 microns	1998 compliant	Offshore, petrochemical piling, maintenance	Low VOC, high build in single coat	Expensive first cost	Unlikely to be used unless cost can be justified by decreasing life cycle costs.
POWDER COATINGS						
Polyester powders and Pvf_2 (refer to Section 5.1.2)	50 microns	Zero	Decorative finish for cladding materials	Effectively solvent free high quality finish	Requires specialist equipment for application	Inappropriate for general fabrication
Polysiloxanes (refer to Section 5.1.4)	125 microns	1988 Compliant	General purpose materials for many applications	Low VOC single coat specifications	Little track record only single supplier	May be more widely used in the future.
HIGH SOLIDS EPOXIES (refer to Sections 5.1.6 and 6.1.2-4)	**75-150 microns**	**1996 and 1998 compliant**	**General purpose materials for all applications**	**VOC compliant, higher film build per coat than conventional materials**	**Need greater care during application than non-compliant materials**	**Currently most widely used compliant coating within the steel fabrication industry.**
WATER-BORNE COATINGS (Refer to Section 5.1.3 and 6.1.2-4)	**75 microns maximum**	**1998 compliant <50g/l**	**General purpose materials for many applications**	**Contain virtually no organic solvent**	**Very sensitive to environmental control during application and curing**	**Unlikely to be widely used in UK general construction in the immediate future**

This CIRIA Report relates to general (ungalvanised) steelwork fabrication for environments normally found in building and civil engineering (*e.g.* specified in accordance with NSSS). It reflects UK legislation, its application in good practice and available materials, as at March 1997. Readers/users should note that the legislation and materials are still being developed, and will change with time.

6 Compliant coatings for construction steelwork

6.1 PRODUCT DEVELOPMENT

The development of compliant materials predates the introduction of any environmental legislation or even the threat of it. Some materials have been in existence – in various forms – for over ten years. The single-coat elastomeric solvent-free urethanes and the hot-applied solvent-free epoxies and urethanes have been available since the 1970s and have been used on bridges for over a decade.

Of more relevance to general construction is the development of high-solids epoxies, whose use can be traced back some 15 years to the introduction of aluminium-pigmented flake epoxies for maintenance painting. When originally launched, environmental awareness was much lower than it is today. Products were developed in direct response to the need to find a replacement for the Type B red lead primer. The material had to be relatively quickly and easily applied, with good film build to surfaces that had received little – or at best minimal – manual surface preparation.

Such has been their success in maintenance applications that they are now very commonly used for this application in a wide range of industrial sectors: all the main construction suppliers have these materials as standard in their product range. Typically, these products have a solids content of 80% and therefore easily comply with legislative requirements.

6.1.1 High solids epoxy coatings

High-solids coatings are similar generally to their low-solids counterparts in respect of application, curing , and the protective properties of the final films. The major differences are the higher viscosity and the higher solids content of the products. Thicker films can be applied, therefore, in fewer coats. Their principal limitations in practice concern application controls (see Section 7.2.3).

The standard definition of 'high-solids' is given in the Glossary. A volume solids percentage level is not stated. Within the coatings industry, however, high-solids coatings are usually volume solids of at least 65%. In most high solids 'compliant' coatings the volume solids is in the range of 65-75%. (A high solids organic solvent-borne compliant coating with a volume solids of 70% would, typically, have a VOC content of 280 g/l).

This CIRIA Report relates to general (ungalvanised) steelwork fabrication for environments normally found in building and civil engineering (*e.g.* specified in accordance with NSSS). It reflects UK legislation, its application in good practice and available materials, as at March 1997. Readers/users should note that the legislation and materials are still being developed, and will change with time.

High solids coatings cannot be formulated simply by reducing the amount of organic solvent. Without other changes, this would only lead to a coating with an unacceptably high viscosity. It is necessary to use low viscosity resins and, in some cases, to add reactive dilutents, which then increase the cross-linking density during curing.

High solids compliant coatings have been formulated successfully by:

- selecting low viscosity, high performance binders (often in conjunction with reactive diluents)
- optimising the solvents used to achieve maximum dilution
- selecting pigments and extenders with low binder demand and optimum particle size distribution
- selecting wetting and dispersing additives carefully to promote good pigment dispersion, high solids content, and low viscosity.

A range of single component (one-pack) and two-component (two-pack, see Glossary) high-solids compliant coatings have been developed with the following general properties:

- high solids
- high film build
- good durability
- good application characteristics
- low temperature curing.

The main emphasis, however, has been on the development of two-pack types.

A comprehensive range of low temperature-curing epoxy, epoxy/acrylic and urethane/acrylic systems have been developed. Typically, epoxy-types have been formulated as primers and intermediate coats with the epoxy/acrylic and urethane/acrylic-types serving as finishes because of their superior UV resistance and durability generally. These two-pack systems have good corrosion resistance and good resistance to abrasion and impact damage.

For most coatings in this class, VOC levels well below the 1996 limits of 400 g/litre for primers and undercoats and 520 g/litre for topcoats have been achieved already, and there is confidence within the coatings industry that the more stringent targets of 250 g/litre and 420 g/litre respectively can be met by 1998.

Examples of two-pack high-solids low VOC compliant coatings are given in Table 4.

High solids epoxies do have some disadvantages and these are considered in Section 7.2.1.

This CIRIA Report relates to general (ungalvanised) steelwork fabrication for environments normally found in building and civil engineering (*e.g.* specified in accordance with NSSS). It reflects UK legislation, its application in good practice and available materials, as at March 1997. Readers/users should note that the legislation and materials are still being developed, and will change with time.

Table 4 Examples of two-pack high solids 'compliant' coatings

Product type	Typical binder type	Typical dry film thickness (microns)	Minimum cure temperature (3 -5)°C	Typical drying and handling times (at 5°C)	Typical volume solids (%)	Typical VOC content (g/litre)
Zinc-rich high-build primer	Two-pack epoxy (low temperature curing)	50-75	0	15 minutes (touch dry) 2 hours (dry to handle)	60	396 (1996 compliant)
High-build zinc phosphate primer	Two-pack epoxy (low temperature)	75-175	0	3 hours (touch dry) 16 hours (dry to handle)	75	243 (1998 compliant)
High-build epoxy MIO intermediate coat/finish	Two-pack epoxy (low temperature curing)	100-150	0	3 hours (touch dry) 16 hours (dry to handle)	75	242 (1998 compliant)
Urethane/acrylic finish	Two-pack epoxy Urethane/Acrylic	50-75	5	8 hours (touch dry) 48 hours (dry to handle)	50	443 (1996 compliant)

Note: Drying times: handling times may vary: reference should be made to manufacturers' literature for specific products.

6.1.2 Water-borne materials (general)

Water-borne coatings represent a significant advance by paint manufacturers towards a safer working environment – not only for the user but also for the manufacturer. The coatings pose a reduced risk of fire, are less odorous, and can be cleaned up more easily. (Water used to clean application equipment after water-borne coatings have been applied should not, however, be discharged directly to waste water systems.) The same is also true of the paints themselves.

Some organic solvent may still be present in water-borne formulations (as a coalescing agent), depending on the type of binder used, but generally, VOC levels are significantly lower than those of solvent-borne 'compliant'coatings and are substantially lower than those of conventional solvent-borne 'non-compliant' coatings. A typical illustration is given in Table 5.

Table 5 Organic solvent released to the atmosphere (1000 square metres coated at a dry film thickness of 200 micrometres)

Conventional coating 'non-compliant'	Solvent-borne compliant coating	Water-borne 'coating'
199 litres of solvent	61 litres of solvent	18 litres of solvent
Assumptions: VOC – 405 g/litre Volume solids – 48%	Assumptions: VOC – 226 g/litre Volume solids – 74%	Assumptions: VOC – 40 g/litre Volume solids – 45%

This CIRIA Report relates to general (ungalvanised) steelwork fabrication for environments normally found in building and civil engineering (*e.g.* specified in accordance with NSSS). It reflects UK legislation, its application in good practice and available materials, as at March 1997. Readers/users should note that the legislation and materials are still being developed, and will change with time.

This great strength is also the Achilles' Heel of the materials. The use of water as the solvent does not alter the fact that, during the curing of the coating, the solvent has to escape from the film. Unlike organic solvents, which are normally volatile and easily escape the curing film, the release of water is very dependent on the immediate environment surrounding the coating during curing.

Curing depends on the air temperature, continuous air movement over the painted component and relative humidity: compared to other forms of epoxy coating, the curing times of water-borne materials are generally much longer in the temperature and humidity ranges normally encountered in a UK fabrication shop.

A minimum substrate temperature of 7°C is usually required to ensure proper film formation. When applied at temperatures below 10°C, drying times are extended significantly and spraying characteristics may be impaired.

Drying times are also significantly increased at relative humidities (RH) in excess of 65%. In conditions of high relative humidity (*i.e.* 80% to 85%), good ventilation, in particular air movement is essential. Relative humidity should not exceed 80% to ensure proper film formation.

For much of the year, the use of these materials could slow down the production process. When this is allied to the time to overcoat, it can become unacceptably long. Generally, water-borne products do not have the same ability to film build as similar solvented materials, and may require additional coats to achieve the same film thickness. Current research may over come these limitations in due course.

Water-borne materials have been successfully used in other countries, notably in the USA (particularly California) and to some extent in Scandinavia. However, both these regions have very different climates from our own, characterised by very long periods of dry weather (as defined by the length of time the relative humidity is below 60%). In the UK, this is not the case: for very long periods of time throughout the year, the RH exceeds this value, making the use of water-borne coatings practically difficult unless special costly provisions are taken to facilitate curing. As practices evolve, of course, this cost disadvantage may lessen.

In designing water-borne compliant coatings, paint manufacturers have sought primarily:

- to maximise the limited application window
- to achieve early water resistance (in applied films)
- to minimise VOC levels
- to maximise surface tolerance
- to achieve durability comparable to solvent-borne counterparts.

The development of water-borne coatings with durability comparable to their solvent-borne counterparts has, thus far, proved challenging. One reason may be that film thicknesses tend to be lower with water-borne formulations (on a coat-for-coat basis).

Almost all types of resins are now available in water-borne versions, including vinyls, acrylics, epoxies, polyesters, alkyds, and urethanes. Each resin has different properties and this has allowed not only one- and two-component primers but also intermediate coats and finishes to be developed.

6.1.3 One-pack water-borne compliant coatings

Product types based on solution polymers and latex-based polymers have been developed.

Solution polymers

Resins for water-borne coatings can be made soluble in water or water-solvent mixtures. They include alkyds and polyesters. Modification of the resin system with polyurethane and acrylic resins is common as the introduction of these higher performance polymeric products improves curing, adhesion, and resistance properties. The resulting coatings dry by evaporation of water and then cure by oxidation.

Water-borne compliant coatings based on this technology have been used most successfully as primers because they have good adhesion to marginally prepared surfaces. They can be applied to steel and aluminium, and can be formulated with minimum film forming temperatures as low as 3°C. The primers are easy to apply, though durability is poorer than latex-based products.

Latex-based polymers

These materials contain high molecular weight polymer particles dispersed in a stabilising aqueous phase. Film formation involves coalescence of the polymer particles as water evaporates from the film.

Latex-based polymers are favoured generally for intermediate coats and finishes. Pure acrylic, styrene/acrylic, polyurethane and polyurethane/acrylic types are common. They may be in thermoplastic or cross-linked form, depending on the hardness and resistance properties desired. Coatings can be formulated with good corrosion resistance, good impact resistance, and good durability. In addition, they can be overcoated easily when maintenance is needed.

Examples of one-pack water-borne compliant coatings are given in Table 6.

6.1.4 Two-pack water-borne compliant coatings

Two-pack water-borne epoxies have been available for several years. They have been used successfully as primers and intermediate coats. Pure epoxy topcoats suffer from chalking on exposure to UV light, so more recent developments have favoured epoxy/acrylic resin systems. Water-borne epoxy/acrylic-based primers, intermediate coats, and finishes are now available.

Water-borne zinc silicate primers have also been available for a long time. Whilst these coatings are environmentally friendly, they can exhibit poorer film forming properties, poorer edge protection, and poorer storage stability at low temperatures, compared with their solvent-borne counterparts.

Examples of two-pack water-borne compliant coatings are given in Table 7.

6.2 THE CURRENT USE OF COMPLIANT MATERIALS

The current use of all forms of compliant coatings has been very limited. Few specifications specifically state that coatings used should meet the solvent limits given in *PG 6/23*. It is this lack of specification that is restricting the use of compliant coatings. It should be noted that for shop applications the fabricator or applicator has a statutory duty to comply with the legislation.

All the main manufacturers have produced technical data sheets to support the launch of their compliant materials; in addition, they have promoted their use both with specifiers and fabricators.

The specifier's reluctant position is, perhaps, understandable. Most are not specially trained or skilled in assessing paint materials or in producing new specifications. A common feeling across the industry is that this can only be addressed by a concerted effort of information dissemination and education.

Table 6 Examples of one-pack water-borne compliant coatings

Product type	Typical binder type	Typical dry film thickness (micrometres)	Minimum film form temperature (°C)	Indicative drying and handling times[(1)] (at 10°C and % RH)	Volume solids (%)	Typical VOC content (g/litre)
Fast drying weldable shop primer	Cross-linking acrylic latex	15-25	5	20 minutes (touch dry) 20 minutes (dry to handle)	25	30-120 (1998 compliant)
Primer/ intermediate coat	Water-soluble modified alkyd or modified acrylic latex	50-75	3-7	2-4 hours (touch dry) 4 hours (dry to handle)	35-40	0-20 (1998 compliant)
MIO intermediate coat/finish	Acrylic/ polyurethene modified acrylic latex	75-100	5-7	2-4 hours (touch dry) 4 hours (dry to handle)	30-35	50-60 (1998 compliant)
Finish	Acrylic/ polyurethane acrylic latex	50-75	5-7	2-4 hours (touch dry) 4 hours (dry to handle)	40-45	50-60 (1998 compliant)

(1) Drying times are dependent on relative humidity and air movement for a given temperature

This CIRIA Report relates to general (ungalvanised) steelwork fabrication for environments normally found in building and civil engineering (*e.g.* specified in accordance with NSSS). It reflects UK legislation, its application in good practice and available materials, as at March 1997. Readers/users should note that the legislation and materials are still being developed, and will change with time.

Table 7 Examples two-pack water-borne coatings

Product type	Typical binder type	Typical dry film thickness (micrometres)	Minimum film form temperature (°C)	Indicative drying and handling times [1] (at 10°C)	Volume solids (%)	Typical VOC content (g/l)
Zinc silicate anti-corrosive primer	Water-borne silicate	50-75	5	60 minutes (touch dry) 2 hours (dry to handle)	46	Nil (1998 compliant)
Epoxy/acrylic primer	Water-borne epoxy amine/acrylic crosslinker	75	5	6 hours (touch dry) 12 hours (dry to handle)	40-45	120 (1998 compliant)
Epoxy/acrylic MIO intermediate coat/finish	Water-borne epoxy amine/acrylic crosslinker	50	5	6 hours (touch dry) 12 hours (dry to handle)	30-35	120 (1998 compliant)
Epoxy/acrylic finish	Water-borne amine/acrylic crosslinker	50	5	6 hours (touch dry) 12 hours (dry to handle)	30-35	120 (1998 compliant)

(1) Drying times will be significantly affected by Relative Humidity and air movement during application and curing.

Away from general construction other sectors of industry are better prepared for the consequences of the EPA:

- The Highways Agency (HA) has now evaluated a range of high-solids materials. These have been accepted and used for more than three years (see Section 9.5.5).
- The UK MOD have approved materials specifications that meet the requirements of *PG 6/23*.
- Most major operators in the oil and gas industries have replacement specifications.

These specifications reflect the particular needs of each industry; for example the oil and gas industry are increasingly using very high solids materials. Here, the improved performance and reduced maintenance more than offset the high initial cost. Such benefits are unlikely to occur in general fabrication.

In the construction industry compliant materials are currently used where a fabricator can get them adopted as an alternative to the issued specification. Invariably, these alternatives make use of the high-solids epoxies. It is still difficult for the fabricator to get these alternatives adopted, for the same reasons that they are not yet specified – the lack of independent and authoritative guidance on the available materials and their likely performance.

Water-borne materials have been successfully used for construction site-based applications in a number of situations:

- British Gas have trialed water-borne materials for maintenance painting.
- The Royal Navy are undertaking trials on operational and museum ships, such as HMS Belfast, on water-borne finishes.

Other materials – *e.g.* solvent-free polyurethanes and high-solids flake materials – are not currently used in general construction. They are however, being increasingly used in other areas, notably in bridge maintenance and for coating vessels and pumps for use in the water industry. The additional expense can only be justified if the improved performance can be shown to reduce maintenance and life-cycle costs.

This CIRIA Report relates to general (ungalvanised) steelwork fabrication for environments normally found in building and civil engineering (*e.g.* specified in accordance with NSSS). It reflects UK legislation, its application in good practice and available materials, as at March 1997. Readers/users should note that the legislation and materials are still being developed, and will change with time.

7 Industry's needs

To understand more fully the benefits of emerging compliant systems, it is necessary to appreciate the needs of the major parties involved. It is then possible to see why their use addresses the concerns of the industry and why the increase in their use makes sound commercial and practical sense, without any loss in performance

7.1 THE COATING MANUFACTURER

All the main manufacturers supplying the construction industry have responded positively to the introduction of the EPA.

With the introduction of the Act, the manufacturers will increase their influence and control over the products that the industry will use. This is because product development has been driven by these coating manufacturers and the limited resin bases available to them. Previously, to a great extent, this was controlled by prevailing standards and specifications, which allowed for a wide range of different resin bases.

The present situation has not arisen because manufacturers have chosen a particular route. Rather, faced with specific formulation targets, they have had a restricted range of options available from raw materials suppliers. Where greater flexibility is allowed, as is the case with finishes, a wider range of compliant materials is available.

7.1.1 Product ranges

Traditional specifications are based on a wide range of different materials. The manufacturer needs to stock – or to be able to make at short notice – a whole spectrum of different materials.

Even a cursory glance at *BS 5493*, reveals the potential scale of stocks. This places an undesirable onus on manufacturers. It makes more commercial sense to have a limited range of products that can cover all the common specification conditions, with a streamlined and more manageable production process. More importantly, this should also be of advantage to the customer and end-user, as they can benefit from the resultant economies of scale and greater standardisation.

Over the last 10 years, manufacturers have moved towards products that attempt more closely to meet the needs of industry. This has led to a situation in which the epoxy-based systems, whether multi-coat or (where appropriate) the primer/finish materials, are predominant in construction. Even where specifications have called for other resin bases, it has become common for an alternative based on epoxies to be proposed.

This CIRIA Report relates to general (ungalvanised) steelwork fabrication for environments normally found in building and civil engineering (*e.g.* specified in accordance with NSSS). It reflects UK legislation, its application in good practice and available materials, as at March 1997. Readers/users should note that the legislation and materials are still being developed, and will change with time.

In general, these have been accepted by the specifiers on the basis of better performance than that required by the specification.

Whilst this makes sense, it needs caution, as in some instances it has been pushed to extremes: examples include the replacement of multi-coat specifications with primer/finishes in external locations and the replacement of, say, multi-coat alkyd schemes with a single coat of epoxy primer. Neither approach is sensible or technically sound.

Manufacturers, in responding to the legislation, in effect are dictating materials by default. It is still possible for specifiers, or fabricators, to influence the materials used: other resin bases, given in current standards and guidance, can still be specified and purchased.

The introduction of the EPA fundamentally changes this position: manufacturers have chosen only two types of coating for primers and build coats to achieve compliance – the water-borne materials and the high solids epoxies. They will have very significant control over the supply of material to the industry, almost independent of what the specifier or fabricator may desire in terms of currently available non-compliant materials.

In addition to this greater control, leading to much greater standardisation, manufacturers have had other reasons for developing the coatings that are now regarded as compliant: their added value. In a conventional material of the late 1980s, something of the order of 50% of 'the product in the can' was organic solvent, serving no useful function once the product was applied (it evaporates). By increasing the solids content, the manufacturers also increase the amount of useful product that is applied to the steel.

These benefits are offset by reduced sales, as less product is sold to achieve the same thickness. At the same time, the coating manufacturers are under pressure to provide the newer materials at the same applied cost (per square metre) as the materials they replace.

7.2 THE FABRICATOR'S NEEDS

Traditionally, the coating of steel to protect against corrosion has been seen as 'a necessary evil', a situation not helped by the lack of standard specifications within the industry and, all too often, over-specification. Both cause the fabricator problems in programme times and application costs and do little to improve quality or service to the customer.

The fundamental problem of conventional materials is the need to apply multi-coats to achieve a given film thickness. In addition, the high VOC contents of many of the materials means that they cannot easily be integrated into the general fabrication process; and more stringent health and safety requirements are increasingly difficult to meet.

7.2.1 Coating application

The application of a coating is often a slow process, – preparation, application curing before over-coating and handling/transportation. These problems have become acute throughout the 1990s, with the increasing trend towards design-and-build construction projects. To lessen delays, fabricators have sought newer materials that reduce the number of applied coats and improve curing times. With conventional materials, there is a limit to what can be achieved, even with the high-build single-coat primer/finishes and the 'winter' type epoxies.

The application of coatings often produces a production bottleneck in the fabrication shop, due to the time delay for curing before overcoating or transportation.

By 1990, the coating industry had launched a new range of epoxy products, based on different resin systems, intended to cure more rapidly and at lower temperatures than their predecessors. If correctly used, these reduced curing times and improved productivity. Even though they were developed before the EPA, these materials were an important step towards compliant coatings, being based on new resin systems with 5-10% greater solids content than their predecessors.

There were other sound commercial reasons for this trend:

- increasing the solids content increased the added value of the product
- the use of high-solids materials rationalised the product ranges, leading to greater standardisation
- by so rationalising, the manufacturers would benefit from resulting economies of scale.

A further development, at approximately the same time, was the introduction of the single-coat specifications based on the so-called 'primer finishes'. These became popular in design-and-build contracts for internal steel in uncontrolled internal environments, such as warehouses. The advantage were:

- a high film build (up to 175 microns) in a single-coat application
- improved productivity and cost reduction by reducing the number of applied coats.

Although the original primer/finishes were not compliant, they did represent a landmark on the way to achieving compliance. For the first time, in the construction industry, there were materials widely available that could easily achieve a nominal thickness in excess of 100 microns that could be used in a typical fabrication shop with relative ease.

These developments in available products all took place prior to the introduction of the EPA in 1990. They occurred as result of commercial judgement and a recognition of the increasing demands from fabricators. Thus, when the EPA was introduced the coating manufacturers were already well placed to develop products that met the requirements of the *PG6/23*.

Despite limitations, the new products did address the practical problems of the fabricator. Compliant materials represent another crucial step (beyond environmental improvement): the need for improved productivity.

A fundamental improvement in using high-solids epoxies that are *PG 6/23* - compliant is the ability to achieve much greater film build in a single coat. It speeds application time. And the higher solids content improves the coverage rate of the material. Their curing times – even at the greater thickness – are also improved, as there is less solvent to escape from the curing film. They have been well received by the fabrication industry.

7.2.2 Health and Safety

It is impossible to provide a complete picture of the modern industry without reference to health and safety. The principal legislation is summarised in Chapters 3 and 4.

In discussing the impact of the EPA with fabricators, other regulations have also been found to be of considerable importance – and have been the primary reasons for fabricators investigating the use of water-borne materials. From the health and safety standpoint, these would be ideal for the fabricator: they avoid many of the problems associated with VOC solvent-borne material. Unfortunately their shortcomings currently outweigh the benefits (see Section 6.1.2).

It must be understood that simple compliance with *PG 6/23* does not imply compliance with other legislation, particularly COSHH and CDM. The EPA is concerned solely with the control of VOCs: while these are health and safety concerns, they are by no means the only harmful constituent of coatings. For example, some of the newer materials may be based on lower molecular weight polymers that can be potentially more hazardous to health than conventional materials. This applies equally to the compliant high-solids epoxies and the water-borne materials. The impact of these would need to form part of a COSHH assessment and the risk will be related to the volume used and the degree of exposure of the operatives.

7.2.3 Application control

The use of high-solids epoxies, in common with all coating applications, is not problem-free. Of obvious and particular concern is the control and achievement of the specified dry film thickness. The application of high solids materials – as opposed to solvent-free does not need specialist new equipment: these materials can be applied using established spraying equipment. However, the operatives applying the paint need

This CIRIA Report relates to general (ungalvanised) steelwork fabrication for environments normally found in building and civil engineering (*e.g.* specified in accordance with NSSS). It reflects UK legislation, its application in good practice and available materials, as at March 1997. Readers/users should note that the legislation and materials are still being developed, and will change with time.

to be trained in the application of the new products. Poor application can result in excessive film build, leading to problems of runs, sags and poorly-cured coatings.

Of equal concern is the problem of under-thickness, particularly where fewer coats are applied. If the minimum dry film thickness is not consistently achieved, then the assumptions regarding the durability of the specification will be invalid and premature failure is highly probable. Again, this problem can be overcome by training of the application operatives to ensure the manufacturers' recommendations are followed.

The control of film thickness is a practical problem associated with the training of operatives and is NOT a fundamental flaw or weakness in the high-solids materials.

Many of the newer materials are also more sensitive to the environmental conditions prevailing at the time of application and during curing. Failure to apply and cure the coatings within the environmental limits stated by the manufacturer can lead to premature failure of the coating. These problems have been encountered in practice.

This is not a fundamental weakness of the new materials, but again highlights the need for the applicator to be familiar with the use of a particular material and the recommendations associated with it, and it is important for the applicator to appreciate differences between materials from different suppliers.

It would be wrong to assume that any of the newer materials can be applied under the same conditions as materials they replace. If the user is unsure then they should check on the product data sheet or confirm the application requirements with the manufacturer directly.

7.3 THE SPECIFIER

The potential advantages are clear for the manufacturer and fabricator of producing and using compliant coatings. For the specifier, they are less clear: they are not immediately or directly affected by the legislation, in the sense that they are not actually operating a proscribed process, nor are they having to produce materials for those who are. However, with change in an inherently conservative industry, the specifier has a key role. The specifier largely controls the products that are used, *via* the project specification. The two main concerns of the specifier are the performance and cost (discussed in Chapters 8 and 9).

The current situation, in which few specifications require compliance, suggests that specifiers are either unaware of the requirements of the EPA, or of the materials available, or are choosing to ignore one or both. The implications of the EPA have

This CIRIA Report relates to general (ungalvanised) steelwork fabrication for environments normally found in building and civil engineering (*e.g.* specified in accordance with NSSS). It reflects UK legislation, its application in good practice and available materials, as at March 1997. Readers/users should note that the legislation and materials are still being developed, and will change with time.

been well broadcast, as has the availability of the materials, not least by the coating manufacturers.

The reluctance to change must be attributed to other reasons, the most obvious being:

- the perceived risk in changing from proven specifications to apparently unproven new materials
- a lack of awareness and experience in modern coatings and corrosion prevention
- obsolete and inappropriate guidance on coatings – *i.e.* materials not specifically aimed at construction
- guidance from the coating suppliers and fabricators not being seen as independent.

In addition to the requirements of the EPA there are good practical reasons for the specifier to initiate change by specifying compliant materials.

A potential advantage for the specifier, in the widespread adoption of compliant materials, is greater standardisation of specifications. The choice of specification of a coating system for any given situation will be much easier and the scope for alternative specifications will be severely restricted.

If the specifier is to have confidence in the selection specification and use of the compliant materials, it is important that there be independent advice on their use. The specifier needs to recognise that the situation is somewhat fluid, in both currently agreed limits on VOC content and available materials. Change may be forced upon the industry or evolve at relatively short notice: the 1996 position is therefore a starting point.

7.4 THE CLIENT

The needs of the client are largely related to two aspects – cost and life expectancy. In general, the client requires minimum first cost and maximum life to first maintenance. These often conflict and compromise is inevitable. In terms of performance, it is important for the client to have confidence in the materials used and to programme inspections and maintenance more accurately.

There is no reason to suppose that the performance of compliant coatings, particularly the high-solids epoxies, will be worse than the materials they replace.

In terms of future maintenance, at the present time, there is no evidence to suggest that it will be any more onerous for compliant coatings. As with traditional coatings, an assessment of maintenance will have to be made, the relevant surface preparation carried out and the structure reinstated with comparable materials.

8 Costing information

8.1 INTRODUCTION

It is impossible here to give definitive cost data on specific products for generic resin bases, even in global terms. Such information is too dependent on many variables peculiar to a given project.

Coating costs are an important concern; it is therefore essential to understand how costs are incurred and where savings might be made in changing from conventional to compliant specifications.

8.2 THE IMPORTANCE OF TOTAL APPLIED COSTS

In considering costs it is important to appreciate that the most important value is the total applied cost, measured as £/m^2.

The importance of this value can be seen from a typical breakdown of costs:

Preparation:	45-50%
Materials:	10-15%
Application time and labour:	35-40%

From this breakdown it is clear that the material cost – although significant – is not the dominant factor.

Assuming that the major cost of preparation cannot be altered, to make significant cost savings in coating operations it is necessary to:

- **reduce the number of coats, and /or**
- **speed up the application and curing times**

Although not always acknowledged, the material cost per unit volume (£/litre) is largely irrelevant in estimating the total cost of providing protection.

8.2.1 Other factors

Cost comparisons (based on £/litre) also take no account of the possibility that the apparently cheaper material may have much poorer coverage (measured in litres/m^2)

This CIRIA Report relates to general (ungalvanised) steelwork fabrication for environments normally found in building and civil engineering (*e.g.* specified in accordance with NSSS). It reflects UK legislation, its application in good practice and available materials, as at March 1997. Readers/users should note that the legislation and materials are still being developed, and will change with time.

than the more expensive material. In addition, the cheaper material may also take longer to cure to overcoat or handle – and may have considerably greater loss factors.

The list prices of materials can also be highly misleading, in that they do not accurately reflect the prices paid by the fabricator. He can attract significant (up to 40%) discounts on the list price, depending on such factors as:

- the buying power and status of the fabricator
- the volume of product purchased over a given trading period.

In evaluating claims that one specification is cheaper (in material terms) than another, it is important to be sure that one is comparing like with like – as opposed to (*e.g.*) the list price of one and the heavily discounted price of the other.

With non-compliant materials, cost reduction is difficult to achieve. These materials have relatively poor film build properties and therefore must be applied in multiple coats to achieve adequate thickness. In construction, the predominant way in which significant savings can be made is to omit decorative finishes, in situations where the steel has no requirement for quality visual appearance.

However, over the last 15 years, there has been a significant reduction in the cost of coating steel in construction: data presented by British Steel indicates a 7% reduction in the costs over the period 1981 to 1991. This was attributed to the recognition that (historically) coatings had been over-specified, particularly with respect to steel used internally.

To produce further savings would need a change in the approach to specifications for corrosion protection. To an extent, the coatings industry has responded by introducing newer products, such as the primer/finishes and faster curing epoxies discussed above.

8.3 FABRICATORS AND OTHER EXPERIENCE

In general, fabricators that have been – or are – keen to adopt high solids compliant coatings are those with an informed view of where the costs of their particular operations lie. They have recognised that the price of a litre of paint is not directly proportional to its final applied costs.

In discussing the use of compliant materials with fabricators as part of the research for this project, it was surprising that they were not forthcoming about the costs of compliant materials. This was not from a reluctance to discuss the issue, but rather that they did not regard it as significant. In general, they felt that, while compliant materials would inevitably attract a premium from the coating supplier, this increase was marginal when compared to the potential benefits in particular:

- improved productivity
- better compliance with Health and Safety legislation.

This recognition is unlikely to be shared by all fabricators and contractors; some will always want to spray on the cheapest available product, irrespective of the real applied costs. Such organisations will change if the trend – influenced by the specifier – encourages them to do so.

8.4 THE COATING APPLICATOR

This Report has referred to 'the fabricator' in a very general manner, making no distinction between the fabricator and the specialist coating applicator. In the context of costs, it is necessary to make such a distinction.

The benefits to a fabricator from using compliant coatings may not necessarily be realised by the specialist applicator:

- The working environment and practices are different.
- The shop is set up solely to apply coatings.
- The applicator is better suited to multi-coat specifications.
- No need to integrate with other operations.
- Reduction in VOC solvent emissions is insufficient to provide benefits.

There is a clear divergence of opinion as to whether the implementation of the EPA will produce any cost savings for coating applicators. The more specialised applicators, who have been forced to change to different materials and are now used to dealing with them, clearly see benefits similar to those available to the general fabricator.

Those applicators who work predominately in the construction industry see the situation very differently. In simple terms, they do not envisage any reduction in costs; in fact, they forecast that costs will increase, in order to maintain the value of their business. Only time and the pressure of market forces will prove which is correct.

8.5 COMPLIANT COATING COSTS

The currently available compliant materials are more expensive (**in terms of £/litre, but not necessarily in £/m^2 applied**) than the materials which they replace. The reasons are openly reported by the manufacturers as:

- the desire – not being achieved – to recover development costs in a short period of time
- the current low sales quantity
- the increase in the volume of solids/litre.

Leaving aside this last point, which will probably mean that the materials will never be the same price per litre as their predecessors, one can expect a decrease in price with time (assuming that other variables, such as the basic resin prices charged by the raw materials suppliers, do not dramatically change).

However, if other materials are specified, such as very high-build solvent-free coatings, this may prove not to be the case. Here the initial applied costs are likely to be greater than for conventional materials and their use could only be based on an assessment of life-cycle costing, where the benefits of improved performance may offset any higher initial cost.

Published cost data for a number of high build coating systems, based on a range of modern materials (such as glass flake pigmented epoxies, vinyl esters and polysiloxanes), show the benefits of reducing the number of coats applied. It is acknowledged that these materials are significantly more expensive (in terms of £/litre) than other available materials.

Recently published data suggests that these systems, using a primer and a single or double coat of high build coating, instead of a 3 or 5 coat conventional system, could produce some very significant savings in applied costs. Estimates from the Armada North Sea gas project suggest reductions in application times from 900 to 600 hours per tonne and a capital saving of 20-25% in applied costs, despite using a more expensive material.

While this example might be regarded as unique, in that it makes use of a new generic material, it illustrates a more general situation in the offshore industry, where there has been an increasing trend to use high build materials in recent times. To a large degree, this has come about as a result of other significant changes within the industry.

8.5.1 The CRINE initiative

The Cost Reduction In the New Era (CRINE) initiative is a wide-ranging examination of all areas of construction in the offshore industry. With respect to coating specifications, it was concluded that there was significant scope for cost savings by changing from traditional multi-coat specifications to newer ones based on higher film build materials. Inevitably, these materials are VOC compliant.

Cost savings are achieved through increases in productivity and a reduction in the life-cycle costs through extending life to first maintenance. These aims have been met by a range of materials including:

- glass flake epoxies and vinyl esters
- solvent-free urethanes
- polysiloxanes
- aluminium metal spray.

It is unlikely that these coatings will find any significant use in general construction; the very high initial cost cannot be justified on the basis of reduction in life-cycle costs.

This CIRIA Report relates to general (ungalvanised) steelwork fabrication for environments normally found in building and civil engineering (*e.g.* specified in accordance with NSSS). It reflects UK legislation, its application in good practice and available materials, as at March 1997. Readers/users should note that the legislation and materials are still being developed, and will change with time.

These changes to coating specifications have resulted from other changes to the working practices and contractual arrangements. The increasing trend towards 'alliancing' and 'partnering' has provided greater scope for bringing about technical change: it has provided an environment in which there is a high degree of collaboration. Such conditions rarely prevail in the majority of construction projects, especially where competitively tendered where there is often insufficient time to make such assessments and, if an alternative is accepted, it is done so at the risk of the specifier.

8.5.2 Bridge construction

A closer comparison with general construction is that of bridges. The Department of Transport specification includes the use of high-solids epoxies that are intended to replace traditional specifications. Typically these specifications are based on three-coat systems, as opposed to between five and seven coats of the older systems. These are expected to produce a reduction in both the initial and life-cycle cost of constructing and maintaining steel bridges.

8.5.3 General construction

In general construction, there is perhaps less scope for significant life-cycle cost savings from the adoption of compliant materials, at least in the case of the high-solids epoxies. Building structures are easier to maintain than either bridges or modules on offshore platforms.

A significant proportion of steel used in general construction has already been coated in epoxy-based materials. The change to compliant coatings would not therefore be expected to produce greatly enhanced durability; and anticipated life-to-first- maintenance would be not be expected to change. The benefits, in terms of cost, are therefore likely to be gained from achieving this performance in fewer applied coats, thereby reducing initial costs.

This CIRIA Report relates to general (ungalvanised) steelwork fabrication for environments normally found in building and civil engineering (*e.g.* specified in accordance with NSSS). It reflects UK legislation, its application in good practice and available materials, as at March 1997. Readers/users should note that the legislation and materials are still being developed, and will change with time.

9 Durability

9.1 GENERAL

The durability of a coating is difficult to define precisely; durability is normally quoted by the specifier in terms of life to first maintenance of the system. The most commonly used lives are those given in *BS 5493*: inherently, there is nothing wrong with them; indeed they have now become so widely-established that it would seem foolish to change them. However, even this document does not simply and clearly define what 'life' actually means, but there is a whole section devoted to how one determines if and when maintenance should be performed.

9.2 LIFE TO FIRST MAINTENANCE

In the specifications included in this Report, the following definition of life to first maintenance applies:

'Life to first maintenance' is the life to the maintenance of the *coating,* not the structure to which it is applied. It does not imply that, after the relevant time period, all the coating must be removed to bare metal and a new system applied. The extent of surface preparation and repainting should be assessed by a detailed survey of the structure after the relevant period.

Life to first maintenance is therefore a compromise. There is little point in doing maintenance if the coating is 100% sound; similarly, there are serious penalties if the coating is left until it has deteriorated to the point where it no longer performs a useful function. Life to maintenance lies between these two extremes and is a matter of engineering judgement. The guidance given in Section 5 of *BS 5493*, although written before the concept of compliant coatings, is still relevant in assessing when maintenance needs to be carried out .

More definitive guidance on life to first maintenance can be, based on a standard level of degradation of the coating system *e.g.* to Ri3 (1% of surface area rusted) as defined in *ISO 4628 Part 3*,

There is sometimes confusion regarding life to first maintenance where a particular structure has a requirement for a high quality decorative finish. The life of this finish, in terms of gloss retention and reflectance, may be less than the desired life to first maintenance for corrosion protection. If this is the case, then decorative maintenance might be needed before the full life of the system is achieved. This judgement is perhaps even more difficult to make than the one regarding general maintenance of the coating as it is subjective to a much greater degree. It can be made less so by undertaking gloss and colour measurement, but will still be a matter of individual judgement.

In the specifications given below in this Report, the life to first maintenance quoted is that for the system in general. Where particular finishes are known to have a lesser decorative life, this is included in the notes for each system.

9.3 FACTORS AFFECTING PERFORMANCE

There are a number of factors that will affect the performance of a coating specification:

- environmental conditions prevailing at the time of application
- the quality of the surface preparation
- the quality of the application
- the dry film thickness
- the physical, chemical and electrical properties of the dry film.

Of these, only the last is affected by whether or not the coating is compliant.

The performance of a coating is inevitably influenced by the specific formulations adopted by a given manufacturer to meet the specified performance. These formulations need to take account not only of the base resin material but also other base components such as hardeners, fillers, curing agents and diluents. Performance will also be influenced by the quality and concentration of the pigment additions to the coating. The manner in which these different components are formulated is a highly complex science that is beyond the scope of this Report.

9.4 BINDERS AND PERFORMANCE

In conventional specifications, the choice of the binding medium had a very significant effect on the performance of a given specification: an epoxy or polyurethane binder would provide much better performance than an alkyd in a similar application. For compliant coatings, the situation is somewhat simplified; the choice is effectively limited to a single generic binding medium – the high-solids epoxies.

In the development of these coatings, the starting point for the manufacturers was to take existing coatings and either remove a proportion of the organic solvent or replace it with a different (less volatile or harmful) one. But this approach has limitations: simply removing the solvent would result in a coating with poor application and storage properties.

For newer materials a more fundamental change is necessary. This retains the generic description of the binder as being an 'epoxy' but has involved the use of a different range of lower molecular weight base resins, curing agents, diluents and less volatile organic solvents. It is only by so doing that the manufacturers have been able to produce *PG 6/23* compliant coatings that also have good storage and application characteristics.

It can therefore be seen that, although the generic description of the binder has remained constant, the component parts have changed. It is these changes that may bring about changes in performance.

This is not a new situation brought about by the advent of compliant coatings. In reality the specifics of coating formulations have always changed over time and generic specifications from different suppliers will always have been based on different specific formulations. These differences in formulation arise because of the cost and availability of raw materials at a given point in time.

It can, therefore, be seen that durability (as defined by life to first maintenance) quoted for specifications based on generic binders is a 'broad brush approach'. Such an approach can provide confidence in generic specification but cannot guarantee that products from different manufacturers or different 'generations', that may be formulated differently, will give identical performance. But they will fall within the specified range.

For example: consider a well-established conventional **non-compliant** generic specification such as:

Primer:	Epoxy Zinc Phosphate	75 Microns Nominal DFT
Intermediate:	Epoxy MIO	75 Microns Nominal DFT

This has been in use for 15-20 years and its performance is well proven in a wide range of environments. The specifier can write a specification and quote a given life with apparent confidence. By implication, the specifier is also **assuming** that the material is identical as that from some time in the past. This is almost certainly not the case: very few formulations will remain constant over time.

The latest formulation will be different to that produced 5 years previously, and may well be radically different from, say, that of 15 years ago; **at the same time, it still generically complies with the specification**. As specifiers are often unaware of these changes, they are readily accepted without demands being made of the supplier to demonstrate 'in-service performance'.

A clear example of such changes occurred in the construction industry in the early 1990s. Specifiers had become familiar with the type of generic specification given above. Around 1990, manufacturers launched a new range of products capable of much faster curing and with the ability to cure to lower temperatures. These were rapidly and widely accepted in the fabrication industry, even though they were not directly specified. They were accepted by the specifiers because they complied with the very general terms of the generic material specification issued, the assumption being that the performance would be the same. In reality, these products were fundamentally different to their predecessors upon which the performance evaluation had been made: **they were all formulated from a new range of resin bases.**

This established practice within the industry, of using generically based specification, will not alter with the use of high solids epoxies that are *PG 6/23* compliant. More importantly, if this approach is accepted, there is no reason to suppose that the newer formulations which are generically the same should not be capable of similar or better performance to their predecessors. From the specifer's and client's perspectives it is important not to establish durability *per se*, rather to have confidence that the changes that may have occurred have not adversely affected the established performance of a generic specification.

9.5 EVALUATING DURABILITY

9.5.1 General

With new products there will always be questions regarding the material's long-term performance and durability. Strictly, the majority of coatings that are now regarded as compliant are not generically new materials; rather, they are developments of existing ones that have been widely used in the construction industry for many years.

In evaluating the durability of compliant coatings, at least with respect to high-solids epoxies, one can therefore start from a position of some confidence that the basic technology is sound.

An increasing number of high-solids materials are formulated differently to the products that they are intended to replace. What is needed is confidence that the changes in formulation have not adversely affected proven durability.

For other materials, such as the water-borne or the polysiloxanes, the situation is different. To a much greater extent, these are generically unknown quantities: they have no immediate predecessors from which likely durability might be extrapolated. A more thorough assessment of likely durability might be needed before they could be considered for a given specific application.

There is a wide range of ways in which the durability of a particular specification can be evaluated. It is unlikely that the full range of evaluation techniques would be needed in any given circumstance. The following methods are described below, with indications of where each method is appropriate:

- track record
- the use of Quality Assured suppliers
- accelerated laboratory testing
- trials and testing by a Third Party

- standard site exposure
- the use of fixed formulations.

9.5.2 Track record

Undoubtedly, the best method of evaluating the durability of a material is to assess its performance, on the basis of use in similar environments and on structures similar to the one for which it is intended to be used. The longer the exposure, the greater the confidence one can have in predicting its future durability. Track record is really the basis, or benchmark, that conventional coatings have come to be judged by; there is a large population of structures, dating back many years, on which judgements can be made.

For apparently new materials, it might be thought (especially by the specifiers) that no track record for a product exists. In certain cases this may be true; for others it is at best a simplification, see example given in Section 9.4.

9.5.3 Manufacturers' Quality Assurance

To have confidence in the claims made for a particular generic specification, the specifier can require that the materials be supplied only from manufacturers that operate an accredited QA system (such as that to *ISO 9000)*. Provided that the accreditation covers design, the manufacturer will have to provide documentary evidence that the performance claims can be justified. Any formulation changes would require similar justification. Inevitably, this cannot be shown through the track record of the new product, but would be demonstrated by benchmark testing against the original material.

Benchmark testing should involve a range of tests to compare the various properties of the coating. For the new formulation to be regarded as appropriate, it would need to perform at least as well as the original. Testing of this sort should be carried out automatically and therefore need not form part of the specification. However, the specification could require that documentary evidence to support performance claims – including the records of independent QA audits – be submitted prior to the material being accepted.

In most cases, a review of this information, in conjunction with the known performance of the generic specification, should provide sufficient confidence in the performance of a given material.

This approach is greatly facilitated by the specifier establishing a working relationship with a number of manufacturers, so that each is aware of the other's needs. Where such a relationship exists, it is appropriate to include a list of preferred suppliers as part of the specification.

This CIRIA Report relates to general (ungalvanised) steelwork fabrication for environments normally found in building and civil engineering (*e.g.* specified in accordance with NSSS). It reflects UK legislation, its application in good practice and available materials, as at March 1997. Readers/users should note that the legislation and materials are still being developed, and will change with time.

9.5.4 Accelerated laboratory testing

Accelerated testing is no substitute for exposure trials and use on real structures, in real naturally-occurring environments. Prediction of actual performance by accelerated tests is extremely difficult. Statistical studies can show a correlation. Such studies need to take account of a large number of variables. This, and the fact that natural weathering is itself an inherently variable process, complicate the prediction of performance for a given application.

The subject of accelerated testing of coatings is a contentious area of paint technology, beyond the scope of this Report. However, it is of use to recognise the advantages and limitations of accelerated testing in the evaluation of a coating's performance.

Accelerated testing is of use for manufacturers in:

- product research and development
- QC testing/bench marking
- comparative testing between products.

Most of the established accelerated tests have been developed over many years for use with conventional coatings. Their use on compliant materials may be inappropriate. A very well-established – almost routine – test for judging the performance of coatings is one of the various salt spray tests that are available. The behaviour of conventional coatings in such a test is well understood. Some compliant coatings perform poorly in this type of test, (*e.g.* the water-borne epoxies and some newer low molecular weight high-solids epoxy zinc-rich primers).

If a conventional epoxy performed poorly in such a test, it would not be expected to perform at all well in natural exposure trials. In the case of the compliant coatings mentioned, the reverse is sometimes found to be the case. This suggests that accelerated test procedures developed for conventional coatings might not always be appropriate for evaluating the newer compliant materials.

It would be difficult to justify a coating specification which required that a given coating should have passed a particular suite of tests to demonstrate that it could achieve a given performance.

If a specifier does need to institute a regime of independent testing for a particular specification, they should be aware of the time and cost involved. Accelerated testing of coatings requires specialist equipment that is not normally found in most materials laboratories; it is usually found in specialist coating test houses or in manufacturers' laboratories. Such equipment is expensive to purchase, maintain and operate.

In addition, coating tests – even accelerated tests – are not quick to carry out. Most tests used to evaluate the long-term performance of coatings are cyclical exposure tests that have to run in excess of 1000 hours (6 weeks). These two factors combine to make testing a relatively expensive and time-consuming exercise. These factors must be considered carefully prior to undertaking any test programme, as should its relevance to the particular application being considered.

9.5.5 Third Party evaluation

In some instances, large specifying authorities or client bodies may themselves carry out both accelerated laboratory testing and site exposure trials of products – examples include the Highways Agency, the MOD and large private sector companies. The results are not normally directly available to the specifier. However, where a particular product passes such testing, it is usually quoted on the relevant manufacturer's data sheet.

Perhaps the widest ranging regime of testing and evaluation is that carried out by the Highways Agency for coatings to be used in road and bridge construction. The details of the testing that is needed are given in the *Specification For Highway Works*. In the HA scheme, materials are tested against an item sheet, if they pass the testing, they are then registered against the Item number. The materials are then permitted to be used for both new and maintenance works. If a specifier seeks the comfort that a rigorous series of test might provide, they can specify that the material has an Item Number.

This method of specifying has disadvantages:

- Manufacturers must supply a detailed formulation for the product that cannot be changed without re-testing.
- Some manufacturers prefer not to reveal formulations, restricting choice.
- Registered paints are sold at a premium to guarantee raw material supply.
- A generically-similar product that is not registered may provide the same performance.

For compliant coatings, a number of products have been tested by various agencies. High-solids epoxies from a number of manufacturers have been tested and approved by the Highways Agency: these are identified by their Item Numbers 111 and 112. These materials now appear in the current edition of the Highway Agency's specification and have been successfully used as alternatives to conventional specifications on a number of projects over the last four years.

9.5.6 BS 5493

BS 5493 is nearly 20 years old and does not cover compliant coatings.

It may be appropriate in some special circumstances to state that a particular product should comply with the Standard's requirements for pigment ratios and contents, provided the paint can also meet the need for compliance with the EPA. If this is done, then the specifier should be aware that this may have a cost implication, both in terms of the material (as the BS places minimum requirements on composition) and more importantly on application. Materials complying with *BS 5493* may have lower film build properties and longer curing times than compliant materials that do not meet the compositional requirements.

This CIRIA Report relates to general (ungalvanised) steelwork fabrication for environments normally found in building and civil engineering (*e.g.* specified in accordance with NSSS). It reflects UK legislation, its application in good practice and available materials, as at March 1997. Readers/users should note that the legislation and materials are still being developed, and will change with time.

9.6 CONCLUSIONS

This Chapter has outlined a number of ways in which a coating's performance can be assessed by the specifier. For general steel construction, none of the methods are ideal, other than for products with a proven track record. For the newer generations of compliant materials, this record will be either non-existent or (at best) recent.

For the majority of construction projects, the specifier should be able to establish sufficient confidence in a particular specification by comparing the test data available from the manufacturers. To some extent this will depend on the relationship between the specifier and the manufacturer.

It is recommended that the specifier establish good working relationships with a number of reputable suppliers, such that a degree of trust and confidence is maintained. If these relationships exist, it is often easy for the specifier to be provided with independent references on a material's previous use. The specifier should also include a list of approved suppliers as part of the specification.

Where such a relationship does not exist, the specifier should consider specifying a material that is known to have been Third Party tested to a particular standard. If the specifier is still in doubt, then they should consider independent testing to demonstrate particular aspects of the materials performance – and these tests should be demonstrably relevant to the particular project.

10 Specifications

The 'CIRIA' specifications given below are based on generic descriptions of materials. They are not based on any particular manufacturer's product range; the full specifications should be available from a wide range of manufacturers. The specifications reflect the current usage of compliant coatings in general fabrication.

The specifications are not directly comparable with the specifications given in *BS 5493,* which does not cover compliant coatings. Nor are the definition of environments within this Standard always appropriate to those found in buildings. It is therefore inappropriate to give direct references to existing specifications found in *BS 5493.*

The assessment of life to first maintenance is based on well-established classifications within the construction industry. The life to first maintenance given for each specification is, at the time of writing, **the minimum anticipated life to first maintenance** as none of the materials given have been used for the periods stated.

The specifications are based on materials that comply with the 1996 requirements of the EPA and generally have been available for a number of years. However, it is probable that many of these materials are now available with 1998-compliance: whilst generically the same as the 1996-compliant formulations they may be compositionally different. Some of the products have been tested by Third Party agencies and some have been trialed on real structures, for up to four years. However, in general the specifications are based on those currently suggested by the major paint manufacturers. For the specifications given, these manufacturers can provide test data comparing the performance of the compliant specifications with the conventional ones they replace.

The material specifications are presented in tabular form, including notes where appropriate. These tables can be readily incorporated into existing specifications, notably the SCI/BCSA *National Steelwork Specification for Building Construction (NSSS),* to provide a stand alone complete specification for tendering.

The specifications are intended for use for new work in general steel construction in the building industry for normal environments. It must be emphasised that they are not intended for more specialist applications, such as:

- highway/rail bridges and similar structures, where particular specifications already exist and should be called up.
- offshore, marine or petrochemical installations: the owners/operators of these installations usually have their own guidance and specifications and these should be consulted.

This CIRIA Report relates to general (ungalvanised) steelwork fabrication for environments normally found in building and civil engineering (*e.g.* specified in accordance with NSSS). It reflects UK legislation, its application in good practice and available materials, as at March 1997. Readers/users should note that the legislation and materials are still being developed, and will change with time.

For the majority of construction projects, it should be possible to have only two coating specifications – or possibly three if the 'no coating' option is taken – to cover the different environments within a building.

However, in some instances, the specifier may be in a position where a greater number are thought to be needed. In such cases, the specifier is advised to try to rationalise the numbers by using the more onerous specification more widely. This may have practical advantages in easing – and even speeding – the production process and removing potential areas of confusion. As such, it may result in more economic fabrication for the contract as a whole. For further clarification, the manufacturer and the fabricator can be consulted.

10.1 COMPATIBILITY WITH FIRE PROTECTION

It is beyond the scope of this document to consider fire protection. However, in many instances, the corrosion protection will be in contact with fire protection applied to the steel. This must be taken into account when writing the coating specification, if compatibility problems are to be avoided.

'Dry board' protection offers no inherent corrosion protection. As such, where this is to be used, the full corrosion protection specification needs to be applied underneath, (with the possible exception of decorative finishes). In this instance, there should be no compatibility problems between the materials, but the specifier should satisfy themselves on this point in the specific contracts.

Cementitious and vermiculite spray fire protection offer no long-term corrosion protection and the full corrosion specification needs to be applied underneath (with the possible exception of decorative finishes). In these instances, the coatings under the fire protection need to have good short-term resistance to alkaline pH. This should be achieved with epoxy-based coatings. If this type of fire protection is to be applied over a zinc-rich epoxy primer, the advice of the coating manufacturer should be sought, as these primers are not alkaline-resistant.

Intumescent paints can form part of the corrosion protection system, provided the advice of the corrosion protection manufacturer is sought. Intumescent paint will usually be organic solvent-based and it may be that these solvents will be incompatible with the corrosion protection. In general, provided that these are epoxy-based, there should not be a problem. However, written confirmation of this compatibility should be obtained from the suppliers of both materials. An easier way of avoiding any potential problems is to specify that the intumescent and corrosion protection are provided by the same manufacturer and to draw attention to the combined use. Written confirmation of compatibility should still be obtained.

The use of intumescents over zinc-rich epoxies needs particular care. A number of cases have been reported where poor adhesion of the intumescent has resulted in premature failure of the intumescent shortly after completion of the structure. The reasons are unclear at the time of writing, although it would appear that the problem is more likely to occur when using a water-borne rather than an organic solvent-borne intumescent. If such a coating system is to be used, the advice of the manufacturers must be sought.

10.2 LIFE TO FIRST MAINTENANCE

For external specifications, the durabilities for three different environments are given for each specification. The environments given are described in terms with which most users will be familiar.

For internal steel coating, specification and durability is less easily defined in such global terms, internal environments being somewhat more variable than those encountered externally. The specifications are therefore based on these commonly encountered in general construction. To assist the specifier in selecting the appropriate specification for a given application, each CIRIA Specification describes the environment.

The approximate equivalent corrosivity categories given in *ISO 12944 Part 2 (draft)* have also been given under each of the environments defined – for information only. A full definition of these categories and the environments they relate to is given in *ISO 12944 Part 2 (draft)*; they are also defined in *ISO 9223: 1992.* The CIRIA specifications which follow represent current good practice in the UK. They do not necessarily concur with the detailed specifications given in *ISO 12944* which covers a wider range of applications than this Report, but would fall within alternative specifications, as permitted by the Standard.

This CIRIA Report relates to general (ungalvanised) steelwork fabrication for environments normally found in building and civil engineering (*e.g.* specified in accordance with NSSS). It reflects UK legislation, its application in good practice and available materials, as at March 1997. Readers/users should note that the legislation and materials are still being developed, and will change with time.

CIRIA SPECIFICATION E-1

Environment: External exposed steel

FOR THIS ENVIRONMENT

Environment	Rural	Urban/industrial	Coastal
Life to first maintenance[1]	25 years	15-20 years	15-20 years
ISO 12944 Environment	C2	C3/C4	C5

USE THIS SPECIFICATION

Surface preparation	Blast clean to Sa2 ½ of *BS7079 Part A1*				
Coat	Material	Thickness[3] (microns)	Maximum VOC/g/litre given in *PG6/23*		Application[2]
			1996	1998 (proposed)	
Primer	Zinc-rich epoxy[7]	75	400	250	Shop
Intermediate	Epoxy MIO	100-125	400	250	Shop
Finish [4,5,6]	Acrylic/Urethane	50	520	420	Site

NOTING

Table Notes

General: The environmental classifications given above do not necessarily concur with those in *ISO12944*. This standard covers a wider range of applications than this Report is intended to cover; moreover, the standard does not necessarily reflect current good practice in UK fabrication. However, the *ISO 12944* draft classifications are also shown – *for information only.*

1. Life to first maintenance is defined in Section 9.2.
2. All materials should be supplied from a single manufacturer and applied in accordance with the manufacturer's instructions.
3. Thickness quoted is NOMINAL dry film thickness (DFT) as defined in *BS5493*.
4. Alternative finishes may be used if particular colour or gloss levels are required. The advice of the coating manufacturer should be sought.
5. This coat may be replaced with 2 coats of water-borne epoxy or acrylic each 50 microns DFT, if prevailing conditions are appropriate. Advice from the manufacturer should be sought. Decorative maintenance may then be required prior to the life to first maintenance.
6. For light colours, the opacity of a single coat may be inadequate to hide the dark grey of some MIO's. In these cases, an additional coat of the finish is recommended.
7. Zinc-rich primers should comply with *BS4652*.

This CIRIA Report relates to general (ungalvanised) steelwork fabrication for environments normally found in building and civil engineering (*e.g.* specified in accordance with NSSS). It reflects UK legislation, its application in good practice and available materials, as at March 1997. Readers/users should note that the legislation and materials are still being developed, and will change with time.

CIRIA SPECIFICATION E-2

Environment: External exposed steel

FOR THIS ENVIRONMENT

Environment	Rural	Urban/Industrial	Coastal
Life to first maintenance[1]	20 years	15-20 years	15-20 years
ISO 12944 Environment	C2	C3/C4	C5

USE THIS SPECIFICATION

Surface preparation	Blast clean to Sa2 ½ of *BS7079 Part A1*				
Coat	Material	Thickness[3] (microns)	Maximum VOC/g/litre given in *PG6/23*		Application[2]
			1996	1998 (proposed)	
Primer	Epoxy zinc phosphate	75	400	250	Shop
Intermediate	Epoxy M1O	100-125	400	250	Shop
Finish [4,5,6]	Acrylic/Urethane	50	520	420	Site

NOTING

Table Notes

General: The environmental classifications given above do not necessarily concur with those in *ISO 12944*. This standard covers a wider range of applications than this Report is intended to cover; moreover, the standard does not necessarily reflect current practice in UK fabrication. However, the *ISO 12944* draft classifications are also shown – *for information only*.

1. Life to first maintenance is defined in Section 9.2.
2. All materials should be supplied from a single manufacturer and applied in accordance with the manufacturer's instructions.
3. Thickness quoted is NOMINAL dry film thickness (DFT) as defined in *BS5493*.
4. Alternative finishes may be used if particular colour or gloss levels are required. The advice of the coating manufacturer should be sought.
5. This coat may be replaced with 2 coats of water-borne epoxy or acrylic, each 50 microns DFT, if prevailing conditions are appropriate. Advice from the manufacturer should be sought. Decorative maintenance may then be required prior to the life to first maintenance.
6. For light colours, the capacity of a single coat may be inadequate to hide the dark grey of some MIO's. In these cases an additional coat of the finish is recommended.

This CIRIA Report relates to general (ungalvanised) steelwork fabrication for environments normally found in building and civil engineering (*e.g.* specified in accordance with NSSS). It reflects UK legislation, its application in good practice and available materials, as at March 1997. Readers/users should note that the legislation and materials are still being developed, and will change with time.

CIRIA SPECIFICATION I-1

Environment: Internal steel

FOR THIS ENVIRONMENT

Environment	Internal steel in controlled or air-conditioned spaces [1]
Life to first maintenance[2]	Life of the building
ISO 12944 Environment	C1

USE THIS SPECIFICATION

Surface preparation	None
Coat Primer Intermediate Finish [3]	No corrosion protection required

NOTING

Table Notes

General: The environmental classifications given above do not necessarily concur with those in *ISO12944*. This standard covers a wider range of applications than this Report is intended to cover; moreover, the standard does not necessarily reflect current practice in UK fabrication. However, the *ISO 12944* draft classifications are also shown – *for information only*.

1. Examples are offices, retail outlets.
2. Life to first maintenance is defined is Section 9.2.
3. If the steel has decorative requirement, then refer to Specification I-2.

This CIRIA Report relates to general (ungalvanised) steelwork fabrication for environments normally found in building and civil engineering (*e.g.* specified in accordance with NSSS). It reflects UK legislation, its application in good practice and available materials, as at March 1997. Readers/users should note that the legislation and materials are still being developed, and will change with time.

CIRIA SPECIFICATION I-2

Environment: Internal steel

FOR THIS ENVIRONMENT

Environment	Internal controlled environment with decorative requirement
Life to first maintenance[1]	Nominally 5-10 years[2]
ISO 12944 Environment	C1

USE THIS SPECIFICATION

Surface preparation	Blast clean to Sa2 ½ of *BS7079 Part A1*				
Coat	Material	Thickness[4] (microns)	Maximum VOC/g/litre given in *PG6/23*		Application[3]
			1996	1998 (proposed)	
Primer	Epoxy zinc phosphate	50	400	250	Shop
Finish [5,6,7]	Acrylic/urethane	50	520	420	Shop

NOTING

Table Notes

General: The environmental classifications given above do not necessarily concur with those in *ISO12944*. This standard covers a wider range of applications than this Report is intended to cover; moreover, the standard does not necessarily reflect current practice in UK fabrication. However, the *ISO 12944* draft classifications are also shown – *for information only*.

1. Life to first maintenance is defined in Section 9.2.
2. This specification is only intended for use where steel has a decorative requirement. Life to maintenance will be determined by decorative needs. In terms of corrosion protection, the life is that of the building. If the steel is hidden, then use Specification I-1.
3. All materials should be from a single manufacturer and applied in accordance with the manufacturer's instructions.
4. Thickness quoted is NOMINAL dry film thickness (DFT) as defined in *BS5493*.
5. The manufacturer should be advised of the required architectural finish in order that the optimum finish can be used. It is important to note that this finish is not needed for corrosion protection and, in the majority of cases, only a single coat should be needed.
6. An alternative specification would be 125 microns DFT of an Epoxy Primer/finish, provided this meets the decorative requirements of the designer/specifier.
7. Alternative finishes may be used if particular colour or gloss levels are required. Coating manufacturers advice should be sought.

This CIRIA Report relates to general (ungalvanised) steelwork fabrication for environments normally found in building and civil engineering (*e.g.* specified in accordance with NSSS). It reflects UK legislation, its application in good practice and available materials, as at March 1997. Readers/users should note that the legislation and materials are still being developed, and will change with time.

CIRIA SPECIFICATION I-3

Environment: Internal steel

FOR THIS ENVIRONMENT

Environment	Steel in cavity walls in clear separation outside the vapour barrier of cladding systems
Life to first maintenance[1]	Life of the building
ISO 12944 Environment	C2

USE THIS SPECIFICATION

Surface preparation	Blast clean to Sa2 ½ of *BS7079 Part A1*				
Coat	Material	Thickness[2] (microns)	Maximum VOC/g/litre given in *PG6/23*		Application[3]
			1996	1998 (proposed)	
Primer	Zinc-rich epoxy [4]	50	400	250	Shop

NOTING

Table Notes

General: The environmental classifications given above do not necessarily concur with those in *ISO12944*. This standard covers a wider range of applications than this Report is intended to cover; moreover, the standard does not necessarily reflect current practice in UK fabrication. However, the *ISO 12944* draft classifications are also shown – *for information only*.

1. Life to first maintenance is defined in Section 9.2.
2. Thickness quoted is NOMINAL dry film thickness (DFT) as defined in *BS5493*.
3. All materials should be from a single manufacturer and applied in accordance with the manufacturer's instructions.
4. Zinc-rich primer should comply with *BS4652*.

CIRIA SPECIFICATION I-4

Environment: Internal exposed steel

FOR THIS ENVIRONMENT

Environment	Semi controlled environments: plant rooms, dry warehouses, roof voids (where occasional condensation may occur)
Life to first maintenance[1]	15-20 years
ISO 12944 Environment	C2

USE THIS SPECIFICATION

Surface preparation	Blast clean to Sa2 ½ of *BS7079 Part A1*				
Coat	Material	Thickness[2] (microns)	Maximum VOC/(g/litre) given in *PG6/23*		Application[3]
			1996	1998 (proposed)	
Primer	Epoxy primer/finish[4]	125	400	250	Shop

NOTING

Table Notes

General: The environmental classifications given above do not necessarily concur with those in *ISO12944*. This standard covers a wider range of applications than this Report is intended to cover; moreover, the standard does not necessarily reflect current practice in UK fabrication. However, the *ISO 12944* draft classifications are also shown – *for information only.*

1. Life to first maintenance is defined in Section 9.2.
2. Thickness quoted is NOMINAL dry film thickness (DFT) as defined in *BS5493*.
3. All materials should be from a single manufacturer and applied in accordance with the manufacturer's instructions.
4. Alternative finishes may be used if particular colour or gloss levels are required. The advice of the coating manufacturer should be sought. An alternative specification where there is no decorative requirement would a single coat of zinc-rich epoxy to 50 microns DFT. If used this should be applied to a surface blast-cleaned to Sa 2½ of *BS7079 Part A1* and the primer should meet the requirements of *BS4652*.

This CIRIA Report relates to general (ungalvanised) steelwork fabrication for environments normally found in building and civil engineering (*e.g.* specified in accordance with NSSS). It reflects UK legislation, its application in good practice and available materials, as at March 1997. Readers/users should note that the legislation and materials are still being developed, and will change with time.

CIRIA SPECIFICATION I-5

Environment: Internal exposed steel - uncontrolled

FOR THIS ENVIRONMENT

Environment	Uncontrolled environments that are frequently damp or wet: swimming pools, kitchens, laundries, etc.
Life to first maintenance[1]	15-20 years
ISO 12944 Environment	C3

USE THIS SPECIFICATION

Surface preparation	Blast clean to Sa2 ½ of *BS7079 Part A1*				
Coat	Material	Thickness[3] (microns)	Maximum VOC/g/litre given in *PG 6/23*		Application[2]
			1996	1998 (proposed)	
Primer	Epoxy zinc phosphate	50	400	250	Shop
Intermediate	Epoxy MIO	125	400	250	Shop
Finish [4,5,6]	Acrylic/urethane	50	520	420	Site

NOTING

Table Notes

General: The environmental classifications given above do not necessarily concur with those in *ISO12944*. This standard covers a wider range of applications than this Report is intended to cover; more over, the standard does not necessarily reflect current practice in UK fabrication.

1. Life to first maintenance is defined in Section 9.2.
2. All materials should be supplied from a single manufacturer and applied in accordance with the manufacturers instructions.
3. Thickness quoted is NOMINAL dry film thickness (DFT) as defined in *BS5493*.
4. Alternative finishes may be used if particular colour or gloss levels are required. The advice of the coating manufacturer should be sought.
5. This coat may be replaced with 2 coats of water-borne epoxy or acrylic each 50 microns DFT, if prevailing conditions are appropriate. Advice from the manufacturer should be sought. Decorative maintenance may then be required prior to the life to first maintenance.
6. For light colours, the opacity of a single coat may be inadequate to hide the dark grey of some MIOs. In these cases an additional coat of the finish is recommended.

This CIRIA Report relates to general (ungalvanised) steelwork fabrication for environments normally found in building and civil engineering (*e.g.* specified in accordance with NSSS). It reflects UK legislation, its application in good practice and available materials, as at March 1997. Readers/users should note that the legislation and materials are still being developed, and will change with time.

References

British Constructional Steelwork Association/Steel Construction Institute. (1994) National structural steelwork specification for building construction. Publication no. 203/94. BCSA/SCI

British Standards Institution. (1977) BS 5493. Code of practice for protective coating of iron and steel structures against corrosion. BSI

British Standards Institution. (1989) BS 7079: Part A1. Specification for rust grades and preparation grades of uncoated steel substrates and of steel substrates after overall removal of previous coatings. BSI

British Standards Institution (1992) BS 2015. Glossary of paint and related terms. BSI

British Standards Institution. (1995) BS 4652 Specification for metallic zinc-rich priming paints (organic media). BSI.

British Standards Institution. (1996) BS EN 971-1. Paints and varnishes – terms and definitions for coating materials. Part 1 – general terms. BSI

British Standards Institution. BS EN ISO 12944-5. Paints and varnishes - corrosion protection of steel structures by protective paint systems. Part 5 - protective paint systems. BSI [Currently being developed].

Construction Industry Research and Information Association. (1993) CIRIA Report 125 - A guide to the control of substances hazardous to health in design and construction. CIRIA.

Control of Pollution Act. (1974) Control of Pollution Act 1974. Chapter 40. HMSO.

Council of the European Communities. (1994) Council Decision 94/904/EEC. Establishing a list of hazardous waste pursuant to 91/689, art.1(4) on hazardous waste. CEC.

Council of the European Communities. (1975) Council Directive 75/442/EEC of 15 July 1975 on waste. CEC.

Council of the European Communities. (1991) Council Directive 91/689/EEC of 12 December 1991 on hazardous waste. CEC.

Council of the European Communities. (1992) Council directive 92/57/EEC of 24 June 1992 on the implementation of minimum safety and health requirements at temporary or mobile construction sites. CEC.

This CIRIA Report relates to general (ungalvanised) steelwork fabrication for environments normally found in building and civil engineering (*e.g.* specified in accordance with NSSS). It reflects UK legislation, its application in good practice and available materials, as at March 1997. Readers/users should note that the legislation and materials are still being developed, and will change with time.

Department of the Environment. (1996) DoE Circular 6/96. Environmental protection act 1990: Part II. Special waste regulations. HMSO.

Department of the Environment. (1991) DoE environmental protection act 1990 - Part 1, GG1(91). Processes prescribed for air pollution control by local authorities. Secretary of State's guidance - Introduction to Part 1 of the act. HMSO.

Department of the Environment. (1991) DoE environmental protection act 1990 - Part 1, GG4(91). Processes prescribed for air pollution control by local authorities. Secretary of State's guidance - Interpretation of terms used in process guidance notes. HMSO.

Department of the Environment. (1976) DoE Waste Management Paper 1. A review of options. A memorandum providing guidance on the options available for waste treatment and disposal. HMSO.

Department of the Environment. (1976) DoE Waste Management Paper 2. Waste disposal surveys. HMSO.

Department of the Environment. (1976) DoE Waste Management Paper 3. Guidelines for the preparation of a waste disposal plan. HMSO.

Department of the Environment. (1994) DoE Waste Management Paper 4. Licensing of waste management. HMSO.

Department of the Environment. (1995) DoE Waste Management Paper 4A. Licensing of metal recycling sites. HMSO.

Department of the Environment. (1976) DoE Waste Management Paper 5. The relationship between waste disposal authorities and private industry. HMSO.

Department of the Environment. (1984) DoE Waste Management Paper 6. Polychlorinated biphenyl (PCB) wastes. A technical memorandum on reclamation, treatment and disposal including a code of practice. HMSO.

Department of the Environment. (1985) DoE Waste Management Paper 7. Mineral oil wastes. A technical memorandum on arisings, treatment and disposal including a code of practice. HMSO.

Department of the Environment. (1985) DoE Waste Management Paper 8. Heat-treatment cyanide wastes. A technical memorandum on arisings, treatment and disposal including a code of practice. HMSO.

Department of the Environment. (1976) DoE Waste Management Paper 9. Halogenated hydrocarbon solvent wastes from cleaning processes. A technical memorandum on reclamation and disposal including a code of practice. HMSO.

Department of the Environment. (1976) DoE Waste Management Paper 10. Local authority waste disposal statistics 1974/5. HMSO.

Department of the Environment. (1983) DoE Waste Management Paper 11. Metal finishing wastes. A technical memorandum on arisings, treatment and disposal including a code of practice. HMSO.

Department of the Environment. (1977) DoE Waste Management Paper 12. Mercury bearing wastes. A technical memorandum on storage, handling, treatment and disposal and recovery of mercury including a code of practice. HMSO.

Department of the Environment. (1977) DoE Waste Management Paper 13. Tarry and distillation wastes and other chemical based residues. A technical memorandum on arisings, treatment and disposal including a code of practice. HMSO

Department of the Environment. (1977) DoE Waste Management Paper 14. Solvent wastes (excluding halogenated hydrocarbons). A technical memorandum on reclamation and disposal. HMSO.

Department of the Environment. (1978) DoE Waste Management Paper 15. Halogenated organic wastes. A technical memorandum on arisings, treatment and disposal. HMSO.

Department of the Environment. (1980) DoE Waste Management Paper 16. Wood-preserving wastes. A technical memorandum on arisings , treatment and disposal including a code of practice. HMSO.

Department of the Environment. (1978) DoE Waste Management Paper 17. Wastes from tanning, leather dressing and fellmongering. A technical memorandum on recovery, treatment and disposal including a code of practice. HMSO.

Department of the Environment. (1984) DoE Waste Management Paper 18. Asbestos wastes. A technical memorandum on arisings and disposal including a code of practice. HMSO.

Department of the Environment. (1978) DoE Waste Management Paper 19. Wastes from the manufacture of pharmaceuticals, toiletries and cosmetics. A technical memorandum of arisings, treatment and disposal including a code of practice. HMSO.

Department of the Environment. (1980) DoE Waste Management Paper 20. Arsenic bearing wastes. A technical memorandum on recovery, treatment and disposal including a code of practice. HMSO.

Department of the Environment. (1980) DoE Waste Management Paper 21. Pesticide wastes. A technical memorandum on arisings and disposal including a code of practice. HMSO.

Department of the Environment. (1980) DoE Waste Management Paper 22. Local authority waste disposal statistics 74/75-77/78. HMSO.

Department of the Environment. (1981) DoE Waste Management Paper 23. Special wastes. A technical memorandum providing guidance on their definition. HMSO

This CIRIA Report relates to general (ungalvanised) steelwork fabrication for environments normally found in building and civil engineering (*e.g.* specified in accordance with NSSS). It reflects UK legislation, its application in good practice and available materials, as at March 1997. Readers/users should note that the legislation and materials are still being developed, and will change with time.

Department of the Environment. (1984) DoE Waste Management Paper 24. Cadmium bearing wastes. HMSO.

Department of the Environment. (1987) DoE Waste Management Paper 25. Clinical wastes. A technical memorandum on arisings, treatment and disposal including a code of practice. HMSO.

Department of the Environment. (1988) DoE Waste Management Paper 26. Landfilling wastes. A technical memorandum on the legislation, assessment and design, development, operation, restoration and disposal of difficult wastes to landfill including the control of landfill gas, economics, a bibliography and glossary of terms. HMSO.

Department of the Environment. (1994) DoE Waste Management Paper 26A. Landfill completion. A technical memorandum. HMSO.

Department of the Environment. (1995) DoE Waste Management Paper 26B. Landfill design, construction and operational practice. HMSO.

Department of the Environment. (1991) DoE Waste Management Paper 27. The control of landfill gas. A technical memorandum on the monitoring and control of landfill gas. HMSO.

Department of the Environment. (1991) DoE Waste Management Paper 28. Recycling. A memorandum providing guidance to local authorities on recycling. HMSO.

Department of the Environment. (1996) Environmental Protection Act 1990. Section 34. Waste Management. The Duty of Care. A Code of Practice. HMSO.

Department of the Environment. (1992) PG6/23(92) Environmental Protection Act 1990, Part1. Secretary of State's guidance - coating of metal and plastic. HMSO.

Department of Transport Highways Agency. (1991) Manual of contract documents for highway works. Volume 1. Specification for highway works, December 1991, reprinted August 1993 with amendments. DTP/HMSO

Environment Act. (1995) Environment act 1995. Chapter 25. HMSO.

Environmental Protection Act. (1990) Environmental protection act 1990. Chapter 43. HMSO.

Health and Safety Commission. (1995) Designing for health and safety in construction. A guide for designers on the Construction (Design and Management) Regulations. HSE Books.

This CIRIA Report relates to general (ungalvanised) steelwork fabrication for environments normally found in building and civil engineering (*e.g.* specified in accordance with NSSS). It reflects UK legislation, its application in good practice and available materials, as at March 1997. Readers/users should note that the legislation and materials are still being developed, and will change with time.

Health and Safety Commission. (1996) L76 Approved supply list (3rd ed.). Information approved for the classification and labelling of substances and preparations dangerous for supply. Chemicals (hazard information and packaging for supply) amendment regulations CHIP 96. HSE Books.

Health and Safety at Work Etc. Act. (1974) Health and safety at work etc. act, 1974. Chapter 37.

Health and Safety Executive. (1994) HSE Environmental Hygiene EH 40/94. Containing the list of maximum exposure limits and occupational exposure standards for use with the control of substances hazardous to health regulations. HMSO

International Organization for Standardization. (1994) ISO 9000. Quality Systems. ISO

International Organization for Standardization. (1992) ISO 9223. Corrosion of metals and alloys - corrosivity of atmospheres; classification. ISO

International Organization for Standardization. (1994) ISO 12944: Part 2 (draft). Paints and varnishes. Corrosion protection of steel structures by protective paint systems. Part 2 classification of environments. ISO

Statutory Instruments. (1980) SI 1980/1709. Public Health, England and Wales; Public Health, Scotland. The Control of Pollution (Special Waste) Regulations. HMSO.

Statutory Instruments. (1988) SI 1988/1657. Health and safety. The Control of Substances Hazardous to Health Regulations. HMSO.

Statutory Instruments. (1991) SI 1991/472. The Environmental Protection (Prescribed Processes and Substances) Regulations. HMSO.

Statutory Instruments. (1991) SI 1991/2431. Health and Safety. The Control of Substances Hazardous to Health (Amendment) Regulations. HMSO.

Statutory Instruments. (1992) SI 1992/614. The Environmental Protection (Prescribed Processes and Substances) (Amendment) Regulations. HMSO.

Statutory Instruments. (1992) SI 1992/2382. Health and Safety. The Control of Substances Hazardous to Health (Amendment) Regulations. HMSO.

Statutory Instruments. (1993) SI 1993/2405. The Environmental Protection (Prescribed Processes and Substances) (Amendment) (No.2) Regulations. HMSO.

Statutory Instruments. (1994) SI 1994/1056. Environmental Protection. The Waste Management Licensing Regulations. HMSO.

Statutory Instruments. (1994) SI 1994/1271. The Environmental Protection (Prescribed Processes and Substances Etc.) (Amendment) Regulations. HMSO.

This CIRIA Report relates to general (ungalvanised) steelwork fabrication for environments normally found in building and civil engineering (*e.g.* specified in accordance with NSSS). It reflects UK legislation, its application in good practice and available materials, as at March 1997. Readers/users should note that the legislation and materials are still being developed, and will change with time.

Statutory Instruments. (1994) SI 1994/1329. The Environmental Protection (Prescribed Processes and Substances Etc.) (Amendment) (No.2) Regulations. HMSO.

Statutory Instruments. (1994) SI 1994/3140. Health and Safety. The Construction (Design and Management) Regulations 1994. HMSO.

Statutory Instruments. (1994) SI 1994/3246. Health and safety. The Control of Substances Hazardous to Health Regulations 1994. HMSO.

Statutory Instruments. (1994) SI 1994/3247. The Chemicals (Hazard Information and Packaging for Supply) Regulations. HMSO.

Statutory Instruments. (1996) SI 1996/972. Environmental Protection. The Special Waste Regulations. HMSO.

Statutory Instruments. (1996) SI 1996/1092. Health and Safety. The Chemicals (Hazard Information and Packaging for Supply) (Amendments) Regulations. HMSO

Statutory Instruments. (1996) SI 1996/2019. Environmental Protection. The Special Waste (Amendment) Regulations. HMSO

Statutory Instruments. (1996) SI 1996/3138. Health and Safety. The Control of Substances Hazardous to Health (Amendment) Regulations. HMSO

Statutory Instruments. (1997) SI 1997/11. Health and Safety. The Control of Substances Hazardous to Health (Amendment) Regulations. HMSO

United Nations Economic Commission for Europe. (1991) Geneva Protocol on the control of volatile organic compound [VOC] emissions or their transboundary fluxes. UNECE.

Biblography

OTHER RELEVANT CIRIA REPORTS

Bradley, R., Griffiths, A. and Levitt, M. (1995) CIRIA Project Report 16. Environmental impact of building and construction materials. Volume F. Paints and coatings, adhesives and sealants. CIRIA.

[One of a set of six reports on the environmental issues affecting a range of building and construction materials]

Construction Industry Research and Information Association. (1993) CIRIA Report 125. A guide to the control of substances hazardous to health in construction. CIRIA.

Construction Industry Research and Information Association. (1995) CIRIA Report 145. CDM regulations - case study guidance for designers: an interim report.CIRIA.

Haigh, I.P. (1982) CIRIA Report 93. Painting steelwork. CIRIA.

UK ENVIRONMENTAL LEGISLATION

Anon.(1995) Croner's Health and Safety Report 19. The construction (design and management) regulations 1994. Croner Publications.

[A comprehensive guide to the CDM Regulations, covering principles, terminology, summary of the main duties, health and safety plan, and health and safety file]

Murley, L. (1996) 1996 Pollution handbook. National Society for Clean Air and Environmental Protection

[A comprehensive guide to UK and European pollution control legislation. National Society for Clean Air and Environmental Protection. It is indexed fully and revised in January of each year.]

Anon. (1996) A resume of the Environmental Act 1995. Clifford Chance.

Health and Safety Executive. (1995) CDM Regulations. How the regulations affect you! HSE.

Health and Safety Executive. (1996) COSHH. The new brief guide for employers. Guidance on the main requirements of the Control of Substances Hazardous to Health (COSHH) regulations 1994. HSE.

This CIRIA Report relates to general (ungalvanised) steelwork fabrication for environments normally found in building and civil engineering (*e.g.* specified in accordance with NSSS). It reflects UK legislation, its application in good practice and available materials, as at March 1997. Readers/users should note that the legislation and materials are still being developed, and will change with time.

Health and Safety Executive. (1993) A complete guide to CHIP. HSE.

Health and Safety Executive. (1995) Designing for health and safety in construction. A guide for designers on the Construction (Design and Management) Regulations 1994. HSE.

Paint Research Association. (1992) PRA Environmental Unit, Special Publication. Environmental Protection Act 1990, Part I - a simple guide for manufacturers of paints, inks and adhesives, industrial finishers, and allied trades. PRA.

Powis, D. (1995) Construction (Design and Management) Regulations 1994. *British Decorator*, September-October, p. 12.

Low VOC compliant coatings

Anon. (1995) Croner's Environmental Special Report 11. Volatile organic compounds. Croner Publications.

Anon. (1994) TechTrends - new developments in paints and organic coatings. International Reports on Advanced Technologies, Innovation 128.

Anon. (1994) Low VOC coatings protect aircraft facility. *Paint and Coatings Industry* [US], 10(10), p. 47.

Appleman, B.R. (1994) Evaluating performance of low VOC coatings: a commentary. *Journal of Protective Coatings and Linings,* 11(5), p. 127-131.

Brevoort, G.H. and Roebuck, A.H. (1991) Selecting cost-effective protective coatings systems. *Materials Performance,* 30(2), p. 39-50.

Brevoort, G.H. and Roebuck, A.H. (1992) Costing considerations for maintenance and new construction work. In: Corrosion 92, National Association of Corrosion Engineers, Paper 335.

Brevoort, G.H. and Roebuck, A.H. (1993) Review and update of the paint and coatings selection guide. *Materials Performance,* 32(4), p. 31- 45.

Brevoort, G.H. (1993) The coatings consumer: understanding initial and long-term painting costs. *Journal of Protective Coatings and Linings,* 10(12), p. 52.

Cornwell, D.W.(1994) Developments on water-based and low VOC coatings for the protection of structural steel. *Surface Coatings International,* 77(7), p. 311.

Gedge, G. (1994) Designer's view of maintenance coating. *Surface Coatings International,* 77(11), p. 462-465.

Gedge, G. (1995) Single-coat specifications: an economic and green solution. *Corrosion Management,* 5, p. 12.

This CIRIA Report relates to general (ungalvanised) steelwork fabrication for environments normally found in building and civil engineering (*e.g.* specified in accordance with NSSS). It reflects UK legislation, its application in good practice and available materials, as at March 1997. Readers/users should note that the legislation and materials are still being developed, and will change with time.

Holm, J. and Stauning, I. (1995) Challenges for environmental policy: learning from front runners. *European Environment,* 5(4), p. 112.

Hudson, R.M. (1987) Environmental factors affecting the use of steel in buildings. In: Corrosion '87, National Association of Corrosion Engineers, p. 425.

Kennedy, R. (1994) Compliance coatings: pros and cons. *European Coatings Journal,* p. 10

Kronborg Nielsen, P. and Holm Hansen, J.(1994) Ecology and economy in the development and use of heavy-duty protective coatings for steel. In: PRA's First Middle East Conference.

Kronborg Nielsen, P. and Holm Hansen, J. (1994) Paint and pollution - a European perspective. *Journal of Protective Coatings and Linings,* 10(7), p. 35.

McMillan, M. (1995) Coatings and VOCs: how the UK industry is responding. *Polymers, Paint and Colour Journal,* 185(4371), p. 7-9.

Mitchell, M. (1994) Reducing solvent content in protective coatings: an overview. *Journal of Protective Coatings and Linings* 11(12), p. 62-69.

Morris, P. (1994) Compliant coatings for exposed steel structures. *Construction Repair,* 8(4), p. 8-10.

Morris, O. P. (1994) Developments in compliant coatings for maintenance of exposed steel structures. *Industrial Corrosion,* 12(3), p. 11.

Nysteen, S. (1994) Practical experience with air-drying water-borne paints for protection of structural steelwork and freight containers. *Surface Coatings International,* 77(7), p. 311-315.

Pratt, E. (1994) VOC compliant protective coatings - the way forward. *Surface Coatings International*, 77(4), p. 132-140.

Storey, D.C. (1995) Environmental legislation and the paint industry - friend or foe? In: Corrosion 95, National Association of Corrosion Engineers.

Williams, D. and Randall, P. (1994) United States Environmental Protection Agency Report EPA/625/R-94/006. Guide to cleaner technologies - organic coating replacements. EPA.

Wood, T. (1995) Recent advances in VOC-compliant technology for protective coatings applications. In: The future of industrial coatings, PRA International Conference, p. 22.

This CIRIA Report relates to general (ungalvanised) steelwork fabrication for environments normally found in building and civil engineering (*e.g.* specified in accordance with NSSS). It reflects UK legislation, its application in good practice and available materials, as at March 1997. Readers/users should note that the legislation and materials are still being developed, and will change with time.

Appendix 1 Review of relevant Standards

Standards are developed nationally (by BSI, in the UK), collaboratively within Europe (by EN), and internationally (by ISO). In addition, ECHOES is involved in the preparation of technical standards.

New standards are often dual-numbered. A new ISO standard may, for example, be adopted by BSI and issued as a BS. CEN may also choose to adopt an ISO standard, where it exists, in order to avoid duplication. Standards, prepared under the auspices of CEN, are issued by BSI as BS EN documents.

Steel standards, prepared by ECISS, are included by BSI in their BS EN series and European pre-standards are issued as BS EN documents. Within BSI, Codes of Practice are now published with BS (and not CP) numbers.

1. EXISTING RELEVANT BRITISH STANDARDS

A small number of existing British Standards for design and workmanship contain sections on the protective coating of construction steelwork.

BS 5493:1977 is a *Code of Practice for protective coating of iron and steel structures against corrosion.* The Standard gives useful guidance on specifications, inspection, maintenance painting, and health and safety. In addition, it defines a number of protective coating schemes for structural steelwork in a range of service environments. Low VOC or 'compliant' coatings are not mentioned, however, as the document is almost 20 years old. *BS 5493* was being revised by the national committee (BDB/7). Parallel activity within ISO (TC35/SC14), however, has overtaken this initiative.

The new ISO Standard which SC14 is producing (*ISO12944*) will, when published, be adopted by EN and will become a European Standard (*EN ISO 12944*). It will also be adopted in the UK as *BS EN 12944*. It is important to note that BS EN 12944 will not replace *BS 5493* totally. Sections of *BS 5493* concerned with galvanising will be replaced by other European Standards, also in development. Part 5 of *ISO 12944* (see 3 below) will supersede the protective coating systems tabulated in *BS 5493*.

BS 5950: – Structural use of steelwork in building, in nine parts is under revision. *Part 1* which gives recommendations for the design of structural steelwork with hot-rolled sections, flats, plates and hollow sections in buildings and allied structures, mentions protective coatings briefly in a section on durability (in its Section 2.5.2). Users of the document are referred, in the text, to *BS 5493* for guidance on adequate methods of protection.

Part 2, which defines materials, and good practice during fabrication and erection, refers briefly to protective treatments in its Sections 4.6 and 5.5. These sections stress

This CIRIA Report relates to general (ungalvanised) steelwork fabrication for environments normally found in building and civil engineering (*e.g.* specified in accordance with NSSS). It reflects UK legislation, its application in good practice and available materials, as at March 1997. Readers/users should note that the legislation and materials are still being developed, and will change with time.

the importance of protecting internal and inaccessible areas well. Specific reference to coating types is not made.

BS 7543:1992 is a guide to durability of buildings and building elements, products and components. The Standard gives guidance on durability, required and predicted service life and design life of (primarily new) buildings and their components parts. The document also gives guidance on presenting information on the service and design life when a detailed brief is being developed.

2. EUROPEAN PRE-STANDARDS

DD ENV 1993 : Eurocode 3. Design of steel structures. One part, *DD ENV 1993-1-1 : 1992* has been published so far. General rules and rules for building (together with a UK National Application Document) give a general basis for the design of buildings and civil engineering works in steel. It also gives detailed rules which are mainly applicable to ordinary buildings. (The National Application document enables it to be used for experimental practical application and comparative studies). A brief reference to protective measures is made in its Section 2.4 (Durability), but without any detail. Reference Standard 10 'Corrosion protection' is called up in its Section B.2.8. This supporting document, when available, will be a European Standard. The existing UK standard called up is *BS 5493.*

ENV 1090-1 : 1995. (This European pre-standard has not, as yet, been issued as a BS ENV document). *Section 10, Protective Treatment,* refers to coatings in its sub-sections 10.1 (General) and 10.3 (Coating methods). Protective treatments are also referred to in its Section 12.6. This document relies solely on *ISO 12944* (to be published) for information on generic types of paint systems and their specification.

3. BS/ISO STANDARDS IN DEVELOPMENT

ISO 12944 Paints and varnishes – Corrosion protection of steel structures by protective coating systems is being developed currently by sub-committee 14 of ISO technical committee 35 (ISO/TC35/SC14). The Convenorship is held by Norway. *ISO 12944* will be in eight parts:

Part 1 General introduction
Part 2 Classification of environments
Part 3 Design considerations
Part 4 Types of surface and surface preparation
Part 5 Protective paint systems
Part 6 Laboratory performance tests methods
Part 7 Execution and supervision of paint work
Part 8 Development of specifications for new work and maintenance

Within *ISO 12944, Parts 1, 2, 5* and 6 have been circulated recently (March 1997) as Final Draft International Standards (FDIS) with a parallel vote in CEN. These should

This CIRIA Report relates to general (ungalvanised) steelwork fabrication for environments normally found in building and civil engineering (*e.g.* specified in accordance with NSSS). It reflects UK legislation, its application in good practice and available materials, as at March 1997. Readers/users should note that the legislation and materials are still being developed, and will change with time.

only be subject to editorial comment, the standards having been circulated as Draft International Standards (DIS) and circulated by BSI as Drafts for Public Comment (DPC) in December 1994. The comments received on these were resolved by the appropriate ISO working groups at their meetings in June 1995.

Part 3 of *ISO 12944* was circulated as a Draft for Public Comment along with *Parts 1, 2, 4 and 6*. Technical changes were made, but were considered to be only minor and this Part will be circulated as a Final Draft Standard.

The remaining parts of *ISO 12944 (Parts 5, 7 and 8)* were issued as Draft International Standards in March 1996 and were circulated as Drafts for Public Comment in April 1996.

Protective paint systems are being considered in *Part 5* of *ISO 12944*. This Part of the Standard will describe different generic types of paints and related products on the basis of their chemical composition and type of film formation process. It will give examples of various protective coating systems that have proved suitable for structures exposed to corrosive stresses and corrosively categories described in *ISO 12944-2*, reflecting current knowledge on a world-wide scale. *ISO 12944-5* will be intended to be read in conjunction with the other parts of *ISO 12944*. The Draft International Standard *ISO 12944-5*, issued on 21 March 1996, defines VOC content of paints in Clause 3.20 and then refers further to low-VOC paint systems in Clause 5.3. The document indicates that paint systems with low-VOC content, designed to meet requirements (not defined) for low emission of solvents are included in the examples of paint systems listed in Annex A (of the document). Annex A is only informative. High-solids paints, water-borne paints – and combinations thereof – are listed. Specific references to VOC contents and to compliant coatings, however, are not included.

4. CEN STANDARDS IN DEVELOPMENT

CEN/TC139/SC1 – Paint systems for steel. Sub-committee 1 of TC139 is considering the documents issued by ISO in respect of the development of *ISO 12944*. The intention is to adopt *ISO 12944*, as a CEN Standard, at an appropriate time. CEN will adopt the various Parts as they are approved so that individual Parts will be CEN Standards before all the Parts have been completed. All members of CEN have been given until the end of 1997 to implement *EN ISO 12944* which means that the UK can retain *BS 5493* until all parts the new CEN Standard are published, as it will result in confusion if sections of *BS 5493* concerned with coatings are replaced by *ISO 12944* a little at a time.

REFERENCES

British Standards Institution. (1977) BS 5493. Code of practice for protective coating of iron and steel structures against corrosion. BSI

This CIRIA Report relates to general (ungalvanised) steelwork fabrication for environments normally found in building and civil engineering (*e.g.* specified in accordance with NSSS). It reflects UK legislation, its application in good practice and available materials, as at March 1997. Readers/users should note that the legislation and materials are still being developed, and will change with time.

British Standards Institution. BS EN ISO 12944. Paints and varnishes – corrosion protection of steel structures by protective paint systems. BSI (Currently being developed).

British Standards Institution. (–) BS 5950. Structural use of steelwork in building. BSI

British Standards Institution. (1990) BS 5950: Part 1. Code of practice for design in simple and continuous construction: hot rolled sections . BSI.

British Standards Institution. (1992) BS 5950 : Part 2. Specification for materials, fabrication and erection : hot rolled sections. BSI

British Standards Institution. (1992) BS 7543. Guide to durability of buildings and building elements, products and components. BSI

British Standards Institution. (–) DD ENV 1993. Eurocode 3. Design of steel structures. BSI

British Standards Institution. (1992) DD ENV 1993-1-1. General rules and rules for buildings (together with United Kingdom National Application Document). BSI.

European Standardisation Committee. (–) ENV 1090. Execution of steel structures. CEN.

European Standardisation Committee. (1995) ENV 1090-1. Rules for building. CEN

European Standardisation Committee. CEN/TC139/SC1. Paint systems for steel. CEN (Currently being developed).

This CIRIA Report relates to general (ungalvanised) steelwork fabrication for environments normally found in building and civil engineering (*e.g.* specified in accordance with NSSS). It reflects UK legislation, its application in good practice and available materials, as at March 1997. Readers/users should note that the legislation and materials are still being developed, and will change with time.

Appendix 2 Environmental legislation in continental Europe

1. EUROPEAN VOC LEGISLATION

Efforts have been underway since 1991 in the EU to develop air emission controls. They appear finally to be reaching a conclusion as the Commission of the European Communities in Brussels has recently published a finalised Proposal for a *Council Directive on Limitation of Emissions of Volatile Organic Compounds due to the use of Organic Solvents in Certain Industrial Activities* (the so-called 'Solvents Directive'). The issue of this document is the first formal step in the European legislative procedure.

In the interim, some member states have developed their own regimes. Others, however, have yet to put any controls in place, preferring to wait for the EU legislation. At the present time, therefore, VOC controls in Europe are still selective, affecting some industries in some countries more than others

National approaches to VOC controls (existing limits and reduction goals) in selected Member States are summarised in Table 8 below.

In this Appendix, progress towards the above *Council Directive* will be reviewed and the VOC legislation in place in selected Member States will be summarised.

1.1 The European 'Solvents Directive'

The proposed *Council Directive on the Limitation of Emissions of Volatile Organic Compounds due to the use of Organic Solvents in Certain Industrial Activities* will apply to users (and manufacturers) of coatings operating under controlled conditions.

First framed in 1991, the document has undergone at least eight unofficial drafts in six years, each one of increasing complexity. Progress was delayed in 1994 when the Directive was placed on hold by the commission while the *Integrated Pollution Prevention Control (IPPC) Directive* (see below) was put on a 'fast track' towards adoption.

The regulation of sixteen sector areas, twelve of which are directly coatings-related, is envisaged in the proposed *Council Directive*. The document will provide for solvent management plans (though a mechanism for implementation has yet to be put forward), VOC thresholds, monitoring and fugitive emission values.

On VOC thresholds, it proposes controls on the VOC emissions of installations operating certain solvent-using processes. In so far as is possible, the intention is to design controls which can be met either by substitution (compliant coatings) or by abatement technology. The *Council Directive* will offer the option to establish a national plan as an alternative to implementing the emission controls specified. This will allow Member States the flexibility to achieve, by other means but on the same time-scale, a similar reduction of emissions as would be achieved by applying the Directive's own emission limits.

1.2 The Integrated Pollution Prevention Control (IPPC) Directive

The *IPPC Directive,* which was adopted finally in September 1996, is far wider in scope than the proposed *'Solvents Directive'*. In an integrated regime, covering releases to air, water and land, and prohibiting pollution from one medium to another, the *IPPC Directive* extends to surface treatments (including paint application) with a solvent consumption *capacity* of more than 150 kilogrammes of per hour or more than 200 tonnes per year. Basing the Directive on consumption capacity and not actual solvent use (output) is punitive and will mean that many more businesses will be subject to integrated control.

The *IPPC Directive* will come into operation three years after its adoption (October 1999). It must be implemented fully by companies eight years thereafter (October 2007).

In the UK, the actual administration of the regime has yet to be decided.

2. VOC LEGISLATION IN SELECTED MEMBER STATES

2.1 Denmark

In Denmark, a Ministry for the Environment was established as long ago as 1971. The first law concerned with the protection of the total environment was issued in 1973. Prior to this, however, there were already in existence some rules on pollution embodied in other laws.

The Danish law on environmental protection lists a large number of industrial activities which are considered to have a high potential of pollution. Paint manufacture and paint application each appear in the listing. Such businesses must have formal approval before they can install new plant or make any changes which will increase emissions to the environment. Approval will only be given if the municipal council is certain that production will not exceed specified limits of emissions to the environment.

Recognising the difficulties faced by local councils in assessing permissible levels of emissions and in ensuring compliance with national policy, a National Agency of Environmental Protection was established. The Agency has published a series of guidelines for the control of air pollution from industrial enterprises, specifying in more detail the general pollution limitation principle expressed in the *Environmental Protection Act*. However, the guidelines are not legally binding and local authorities can deviate from the emission limits given and from the list of classified installations. In addition, the guidelines are substance-orientated and do not fix emission limits for

This CIRIA Report relates to general (ungalvanised) steelwork fabrication for environments normally found in building and civil engineering (*e.g.* specified in accordance with NSSS). It reflects UK legislation, its application in good practice and available materials, as at March 1997. Readers/users should note that the legislation and materials are still being developed, and will change with time.

specific industrial or commercial activities. The guideline on the limitation of emissions to the air (Number 6) stresses the necessity for preventive measures in the following terms:

> Polluting materials should be replaced with non-polluting (or less polluting) materials wherever possible .
> Manufacturing methods and equipment must be arranged to minimise emissions to the air .
>
> If necessary, air pollution must be limited by the use of clean technologies. Such technologies should be selected on the basis of technical and economical feasibility (effectively BATNEEC). The competitive position of any company should not be weakened.
> Unavoidable emissions to air must be arranged in such a way that they are not a hazard to health, not a nuisance to human beings and not a source of damage to either plants or animals.

Guideline Number 6, which is similar to that of the German *TA Luft*, groups substances into three classes (I, II and III). Emission limits for volatile organic solvents in paints, however, are considerably lower in the Danish document.

The Danish *Environmental Protection Act* was strengthened in 1991. New items were introduced into the existing legislation, including 'the principle of cleaner technology'. In paints, this involved the use of environmentally-friendly raw materials and the application of coatings with little or no organic solvents. In support of these amendments, binding agreements with groups of paint manufacturers were established on quota for VOC reductions. This initiative is related to the Dutch KWS 2000 project.

2.2 France

In France, no legally binding and generally applicable emission limits for the whole sector of industrial installations using organic solvents are laid down. For installations which emit more than 500 kg/day, however, continuous monitoring is required. It is, nevertheless, the intention of the State Department of Environment to reduce VOC emissions by 30% between 1985 and 2000. Towards this goal, an agreement was signed in February 1986 between the French Government and the French Paint and Varnish Industry Association (FIPEC). Under this convention, it was agreed that the average organic solvent content of paints would be reduced by 25% within five years. This reduction would, in effect be achieved by increasing the production of water-borne paints, high-solids paints, and powder coatings.

At the present time, the concentration of VOCs must not exceed 150 mg/m^3 of air discharged. National legislation is in preparation.

2.3 Germany

In Germany, a voluntary agreement to reduce organic solvent usage by 20-25% between 1983 and 1989 was reached between the Government and the Paint Producers' Association. During this time, however, there was in fact an increase in solvent usage, due to a general increase in market volumes. As a consequence, the voluntary agreement was replaced by a *Technical Directive for the Prevention of Air Pollution*, known as *TA Luft*.

TA Luft has the legal status of an administrative directive and is binding. It groups organic compounds into one of three Classes (Classes I to III) and lays down a series of VOC emission limits (varying form 0.1 to 150 mg/m^3). Typical solvents used in paints are usually blends of organic compounds found in Classes II and III.

In the original *TA Luft Directive*, a paint usage criterion of 250 kg/hour was also defined. Applicators who exceeded this limit were required to obtain formal approval. This approval limit was amended, however, in 1986 from kg/hour of *paint* to kg/hour of *solvent*. The re-definition later favoured applicators who changed to high-solids coatings or water-borne coatings.

2.4 Italy

In Italy, regional restrictions in the most highly industrialised areas of Italy, on the emission of VOCs in the 1980s led to a national law on emissions in 1988 *(DPR203/88)*. National guidelines were then issued on VOC emission limits in 1990 *(Decree N51)*, though these guidelines related only to existing plants. The national guidelines follow the German *TA Luft Directive* and emissions are limited within ranges (0.1 to 600 mg/m^3), depending on the class of the volatile organic compound.

2.5 The Netherlands

In The Netherlands, there are several Acts concerned with the prevention of environmental pollution. One of these Acts is concerned with preventing air pollution *(Air Pollution Prevention Act)*. As yet, however, formal legislation to reduce the VOC emissions is not in place.

The country has a *National Environment Plan* and a voluntary agreement to reduce VOC emissions significantly by the year 2000 has been reached between the Dutch Administration (specifically the Department of Housing, Planning and Environment) and the appropriate Trade Associations. Under Project KWS 2000, VOC emissions from paint application (and other industrial activities in which organic solvents are used) will be reduced by 50% during the period 1981 to 2000. This figure of 50% is, however, an overall target. Taking into account increased production (volume sales) during the period, the specific target reduction of VOCs for industrial painting has been set at 57%.

2.6 Switzerland

Swiss VOC legislation is the most strict of all the European countries. Proposed legislation could reduce VOC emissions to one tenth of those proposed in Germany under *TA Luft*. The Swiss have been very successful at reducing the organic solvent content of paints in the last 35 years through voluntary agreement and changes in the law. In the period 1960 to 1987, for example, paint production increased by 224% but solvent consumption increased by only 33%. This equates with a relative reduction of solvent used (per tonne of paint) of more than 60%.

In 1985, the Swiss Federal Office and the Swiss Paintmakers' Association reached agreement to reduce further the organic solvent of paint by 25% over the following five years; a 12% reduction was claimed after three years.

In 1992, it was proposed that *Swiss Law on the Protection of the Environment* should be amended to introduce a special tax on organic solvents. It has been indicated recently that the tax will be included in proposed revisions to *Swiss law* on environmental protection. When the revised law will be published and enter force is not, as yet, clear. It has been disclosed, however, that the tax will apply not only to solvents but also to solvent-containing preparations, such as paints.

Table 8 Extent of national VOC controls and reduction goals in selected European Countries

Country	Limit	Control methods	Reduction goals
Denmark	Up to 300 mg/m^3	Emission limit related to mass flow	
France	Up to 150 mg/m^3	Emission limit for selected installations, eg core coating application	By 30% by 2000, based on 1980
Germany	Up to 150 mg/m^3	Emission limit related to mass flow (TA Luft)	By 30% by 1999, based on 1988
Italy	20-600 mg/m^3 maximum 50 mg carbon/m^3	Emission limit related to mas flow	
Netherlands		KWS 2000 voluntary industry/government agreement	By 50% by 2000, based on 1980
Sweden			By 50% by 2000, based on 1988 plus action programmes
Switzerland			By 50% by 1995, based on 1987

REFERENCES

Commission of the European Communities. (1996) COM(96) 538 final. Proposal for a council directive on limitation of emissions of volatile organic compounds due to the use of organic solvents in certain industrial activities (submitted by the Commission on 18 February 1997). Official Journal of the European Communities (C99/32) 26 March 1997. CEC.

Council of the European Communities. (1996) Council directive 96/61/EC of 24 September 1996 on integrated pollution prevention and control. CEC.

Danish Environmental Protection Act (1973 and 1991) (Act Number 358).

This CIRIA Report relates to general (ungalvanised) steelwork fabrication for environments normally found in building and civil engineering (*e.g.* specified in accordance with NSSS). It reflects UK legislation, its application in good practice and available materials, as at March 1997. Readers/users should note that the legislation and materials are still being developed, and will change with time.

Netherlands National Environment Plan. (1989) Voluntary agreement to reduce VOC emissions from paint application (Project KWS 2000).

Swiss Federal Government. (1985) Ordinance on air pollution control.

BIBLIOGRAPHY

Allemand, N (1990) *Legislation on VOC Emission Reduction in France and Objectives of the CEC, PRA International Conference, 'Paint and the Environment'*, Copenhagen, Paper 6.

Chatton, P (1995) EU Regulations and their Impact on the Coatings Industry, *Surface Coatings International*, 78 (11), 486. A general survey covering progress on the existing chemicals programme and the VOC directive, with special reference to coatings.

Doorgest, T (1990) *Environmental Regulations in The Netherlands*, PRA International Conference, 'Paint and the Environment', Copenhagen, Paper 5.

Jotischky, H (1995) Coatings under the Legislative Spectre, PRA's Fifteenth International Conference, Brussels.

Meller, E (1991) Current Trends in Environmental Protection Legislation, *Oberflaeche/JOT*, 31 (9), 86.

Rasmussen, J (1990) *Environmental Regulations in Denmark,* PRA International Conference, 'Paint and the Environment', Copenhagen, Paper 4.

Appendix 3 Environmental legislation in the USA

1. INTRODUCTION

For the last 30 years, America has pioneered increasingly stringent air pollution legislation. *Rule 66* of Los Angeles County set the precedent for the first federal legislation in the *Air Quality Act* of 1967. Subsequent amendments of this Act have led to the current *Clean Air Acts* (CAA) of 1970, 1977, and 1990. The 1990 Act established an elaborate regime, dependent for its operation on local, regional, technical and economic circumstances.

The degree to which manufacturers and users of protective coatings are regulated depends largely on locality. Industrial areas, for example, are required to comply with this legislation and reduce air pollution. Limits are being set for maximum VOC contents of specific coating types and many painting shops are now subjected to daily or yearly quota of total VOC emissions.

2. THE AMERICAN APPROACH TO ENVIRONMENTAL LEGISLATION

The American approach is based on:

- two-pronged control targets, regulating ambient emissions to the atmosphere at large, as well as hazardous air pollutants (HAPs) in the immediate work environment.
- a dual set of regulatory controls with responsibilities divided between the federal Government and the individual States in some cases operating independently, in other cases overlapping.
- an overall regulator, the Environmental Protection Agency, with double rule-making powers: to establish model emission guidelines for state implementation in the case of ambient air, but also to set nationwide rules for uniform application in the case of HAPs.
- a facility for arriving at consensus rules through the so-called 'reg-neg process', in which manufacturers (represented by the National Paint and Coatings Association (NPCA)), users, environmental groups and the local community all participate

The Environmental Protection Agency has defined a number of coating types, including High Performance Coatings and Architectural and Industrial Maintenance (AIM) Coatings. The inclusion of these two categories has caused some confusion as there is considerable overlap. The issue has been clouded further by the introduction of a sub-category (Rust-preventative Coatings) as part of High Performance Coatings.

This CIRIA Report relates to general (ungalvanised) steelwork fabrication for environments normally found in building and civil engineering (*e.g.* specified in accordance with NSSS). It reflects UK legislation, its application in good practice and available materials, as at March 1997. Readers/users should note that the legislation and materials are still being developed, and will change with time.

The only difference between High Performance and Industrial Maintenance appears to be the intended use. More emphasis seems to have been placed on AIM coatings in negotiations on VOC limits.

In order to describe the present, complex situation in the USA, in respect of AIM coatings, it is necessary to understand the roles of the main players and the positions which they have adopted.

3. THE REGULATORY PROCESS

Congress provides the general legislative framework for air pollution control (the CAA of 1990). This is then implemented by individual States. The Environmental Protection Agency is the primary regulatory authority, with responsibility for two sets of controls – setting general quality objectives for ambient air pollutants via National Air Quality Standards (NAAQs) and issuing uniform *National Emission Standards for Hazardous Air Pollutants* (NESHAPs). NAAQs serve as the pattern for individual state implementation. NESHAPs apply to both new and existing pollutant sources.

The Environmental Protection Agency is also authorised to review and approve state implementation plans. In cases of disagreement, the Agency can impose its own *Federal Implementation Plan* (FIP). Individual States are responsible for preparing *State Implementation Plans* (SIPs) to attain and maintain each of the NAAQs set by the Environmental Protection Agency. States also have greater control over emissions from existing (emission) sources than new ones. States – or indeed localities within a State – can impose their own stricter emission standards than the minimum standards set at national level (an approach pioneered by the South Coast Air Quality Management District in California).

The NPCA, as the nationwide trade association of the coatings industry, is the chief lobbyist in the consensus process. It has devised its own VOC plan for AIM coatings, with slightly lower reduction targets than those set by the Environmental Protection Agency. The NPCA has been active recently in forestalling the AIM VOC limits which the Agency is seeking to impose on US industry.

4. A NATIONAL VOC STANDARD FOR AIM COATINGS

In the Summer of 1994, the three year-long 'reg-neg' discussions on AIM coatings broke down. As a result, the initiative for developing a VOC standard reverted to the EPA. As an interim measure, NPCA issued a 'model rule' for VOC limits in 1995 (effective in 1996). For rust-preventive coatings, for example, a VOC limit of 400 g/l was suggested.

Some States took this guidance and finalised rules for VOC limits themselves. In Kentucky, Massachusetts, Oregon and Rhode Island, for example, the VOC limit on AIM coatings was set at 450 g/l. Indiana, however, set the VOC limit at 350 g/l. Many other States suspended activity on setting VOC limits in anticipation of an EPA rule.

On 25 June 1996, the long awaited, proposed rule on the VOC content of AIM coatings was published by EPA in the Federal Register. Responses were required by 4 November 1996. The proposed rule will take effect from April 1997, according to the Federal Register. NPCA have indicated, however, that this date may well be extended to January 1998.

The proposed rule sets VOC limits fifty five categories of AIM coatings, including : Industrial Maintenance Coatings (450 g/l), Metallic Pigmented Coatings (500 g/l) Rust Preventive Coatings (400 g/l) and Pretreatment Wash Primers (780 g/l). When effective, it is estimated that the rule will reduce total emissions of VOCs from paints and coatings in the US by an estimated 96,000 tonnes.

States will be required to adopt VOC regulations at least as strict as those imposed by the proposed rule. States such as California, however, which has been regulating the VOC content of AIM coatings for twenty years, may well adopt stricter levels.

In the earlier (1994) non-consensus draft of the proposed rule, EPA proposed a second level of standards for the year 2000 and beyond. This second level of standards was not received well by industry and has been withdrawn. EPA now proposes a second study with industry to establish the need for, and the viability of, further reductions in VOC levels.

5. POSTSCRIPT : DEREGULATION

A fundamental shift in philosophy by Congress in 1995 may call into question some previous legislation, including the *Clean Air Act*. The introduction of common-sense and accountability into the regulatory system appears to be the driving force. One of eight new bills introduced into the House of Representatives seeks to repeal all provisions of the *Clean Air Act Amendment 1990*. The elaborate regime of US air pollution controls described above may well be standing at a crossroads.

REFERENCES

Regulation 66 of Los Angeles County. (1966) Rule 66 (enacted in 1967). Emission limits for photochemically reactive hydrocarbons and oxidants. State of California, USA.

Air Quality Act. (1967) (Federal Law).

Clean Air Act (CAA). 1970. USA (Federal Law).

Clean Air Act Amendment (CAAA). (1977). USA (Federal Law).

Clean Air Act Amendment (CAAA). (1990). USA (Federal Law).

This CIRIA Report relates to general (ungalvanised) steelwork fabrication for environments normally found in building and civil engineering (*e.g.* specified in accordance with NSSS). It reflects UK legislation, its application in good practice and available materials, as at March 1997. Readers/users should note that the legislation and materials are still being developed, and will change with time.

BIBLIOGRAPHY

Appleman, B R (1994) Development of a National VOC Rule : A Review and Update, *Journal of Protective Coatings and Linings*, 11 (10), 19. A comprehensive review of progress towards a national regulation for VOC levels for architectural and maintenance coatings

Jotischky, H (1995) VOC Control at the Cross-Roads : AIM Coatings, *PRA SHE Alert*, 7, 291.

This CIRIA Report relates to general (ungalvanised) steelwork fabrication for environments normally found in building and civil engineering (*e.g.* specified in accordance with NSSS). It reflects UK legislation, its application in good practice and available materials, as at March 1997. Readers/users should note that the legislation and materials are still being developed, and will change with time.